INTERNATIONAL SERIES IN **MATHEMATICS**

BASIC REAL ANALYSIS

The Jones and Bartlett Publishers Series in Mathematics

Geometry

Geometry with an Introduction to Cosmic Topology
Hitchman (978-0-7637-5457-0) © 2009

Euclidean and Transformational Geometry: A Deductive Inquiry
Libeskind (978-0-7637-4366-6) © 2008

A Gateway to Modern Geometry: The Poincaré Half-Plane, Second Edition
Stahl (978-0-7637-5381-8) © 2008

Understanding Modern Mathematics
Stahl (978-0-7637-3401-5) © 2007

Lebesgue Integration on Euclidean Space, Revised Edition
Jones (978-0-7637-1708-7) © 2001

Precalculus

Precalculus: A Functional Approach to Graphing and Problem Solving, Sixth Edition
Smith (978-0-7637-5177-7) © 2010

Precalculus with Calculus Previews (Expanded Volume), Fourth Edition
Zill/Dewar (978-0-7637-6631-3) © 2010

Precalculus with Calculus Previews (Essentials Version), Fourth Edition
Zill/Dewar (978-0-7637-3779-5) © 2007

Calculus

Calculus of a Single Variable: Early Transcendentals, Fourth Edition
Zill/Wright (978-0-7637-4965-1) © 2010

Multivariable Calculus, Fourth Edition
Zill/Wright (978-0-7637-4966-8) © 2010

Calculus: Early Transcendentals, Fourth Edition
Zill/Wright (978-0-7637-5995-7) © 2010

Multivariable Calculus
Damiano/Freije (978-0-7637-8247-4) © 2011

Calculus: The Language of Change
Cohen/Henle (978-0-7637-2947-9) © 2005

Applied Calculus for Scientists and Engineers
Blume (978-0-7637-2877-9) © 2005

Calculus: Labs for Mathematica
O'Connor (978-0-7637-3425-1) © 2005

Calculus: Labs for MATLAB
O'Connor (978-0-7637-3426-8) © 2005

Linear Algebra

Linear Algebra with Applications, Seventh Edition
Williams (978-0-7637-8248-1) © 2011

Linear Algebra with Applications, Alternate Seventh Edition
Williams (978-0-7637-8249-8) © 2011

Linear Algebra: Theory and Applications
Cheney/Kincaid (978-0-7637-5020-6) © 2009

Advanced Engineering Mathematics

Advanced Engineering Mathematics, Fourth Edition
Zill/Wright (978-0-7637-7966-5) © 2011

An Elementary Course in Partial Differential Equations, Second Edition
Amaranath (978-0-7637-6244-5) © 2009

Complex Analysis

A First Course in Complex Analysis with Applications, Second Edition
Zill/Shanahan (978-0-7637-5772-4) © 2009

Complex Analysis for Mathematics and Engineering, Fifth Edition
Mathews/Howell (978-0-7637-3748-1) © 2006

Classical Complex Analysis
Hahn (978-0-8672-0494-0) © 1996

Real Analysis

An Introduction to Analysis, Second Edition
Bilodeau/Thie/Keough (978-0-7637-7492-9) © 2010

Basic Real Analysis
Howland (978-0-7637-7318-2) © 2010

Closer and Closer: Introducing Real Analysis
Schumacher (978-0-7637-3593-7) © 2008

The Way of Analysis, Revised Edition
Strichartz (978-0-7637-1497-0) © 2000

Topology

Foundations of Topology, Second Edition
Patty (978-0-7637-4234-8) © 2009

Discrete Math and Logic

Discrete Structures, Logic, and Computability, Third Edition
Hein (978-0-7637-7206-2) © 2010

Essentials of Discrete Mathematics
Hunter (978-0-7637-4892-0) © 2009

Logic, Sets, and Recursion, Second Edition
Causey (978-0-7637-3784-9) © 2006

Numerical Methods

Numerical Mathematics
Grasselli/Pelinovsky (978-0-7637-3767-2) © 2008

Exploring Numerical Methods: An Introduction to Scientific Computing Using MATLAB
Linz (978-0-7637-1499-4) © 2003

Advanced Mathematics

Mathematical Modeling with Excel
Albright (978-0-7637-6566-8) © 2010

Clinical Statistics: Introducing Clinical Trials, Survival Analysis, and Longitudinal Data Analysis
Korosteleva (978-0-7637-5850-9) © 2009

Harmonic Analysis: A Gentle Introduction
DeVito (978-0-7637-3893-8) © 2007

Beginning Number Theory, Second Edition
Robbins (978-0-7637-3768-9) © 2006

A Gateway to Higher Mathematics
Goodfriend (978-0-7637-2733-8) © 2006

For more information on this series and its titles, please visit us online at http://www.jbpub.com/math. Qualified instructors, contact your Publisher's Representative at 1-800-832-0034 or info@jbpub.com to request review copies for course consideration.

The Jones and Bartlett Publishers International Series in Mathematics

Mathematical Modeling with Excel
Albright (978-0-7637-6566-8) © 2010

An Introduction to Analysis, Second Edition
Bilodeau/Thie/Keough (978-0-7637-7492-9) © 2010

Basic Real Analysis
Howland (978-0-7637-7318-2) © 2010

Advanced Engineering Mathematics, Fourth Edition, International Version
Zill/Wright (978-0-7637-7994-8) © 2011

Multivariable Calculus
Damiano/Freije (978-0-7637-8247-4) © 2011

Real Analysis
Denlinger (979-0-7637-7947-4) © 2011

Mathematical Modeling for the Scientific Method
Pravica (978-0-7637-7946-7) © 2011

A Journey into Partial Differential Equations
Bray (978-0-7637-7256-7) © 2011

Functions of Mathematics in Liberal Arts
Johnson (978-0-7637-8116-3) © 2012

For more information on this series and its titles, please visit us online at http://www.jbpub.com/math. Qualified instructors, contact your Publisher's Representative at 1-800-832-0034 or info@jbpub.com to request review copies for course consideration.

INTERNATIONAL SERIES IN MATHEMATICS

BASIC REAL ANALYSIS

James S. Howland
University of Virginia

JONES AND BARTLETT PUBLISHERS
Sudbury, Massachusetts
BOSTON   TORONTO   LONDON   SINGAPORE

World Headquarters
Jones and Bartlett Publishers
40 Tall Pine Drive
Sudbury, MA 01776
978-443-5000
info@jbpub.com
www.jbpub.com

Jones and Bartlett Publishers
Canada
6339 Ormindale Way
Mississauga, Ontario L5V 1J2
Canada

Jones and Bartlett Publishers
International
Barb House, Barb Mews
London W6 7PA
United Kingdom

Jones and Bartlett's books and products are available through most bookstores and online booksellers. To contact Jones and Bartlett Publishers directly, call 800-832-0034, fax 978-443-8000, or visit our website, www.jbpub.com.

> Substantial discounts on bulk quantities of Jones and Bartlett's publications are available to corporations, professional associations, and other qualified organizations. For details and specific discount information, contact the special sales department at Jones and Bartlett via the above contact information or send an email to specialsales@jbpub.com.

Copyright © 2010 by Jones and Bartlett Publishers, LLC

All rights reserved. No part of the material protected by this copyright may be reproduced or utilized in any form, electronic or mechanical, including photocopying, recording, or by any information storage and retrieval system, without written permission from the copyright owner.

Production Credits:
Publisher: David Pallai
Acquisitions Editor: Timothy Anderson
Editorial Assistant: Melissa Potter
Production Director: Amy Rose
Senior Marketing Manager: Andrea DeFronzo
V.P., Manufacturing and Inventory Control: Therese Connell
Composition: Northeast Compositors, Inc.
Cover and Title Page Design: Kristin E. Parker
Cover Image: © Comstock Images/age fotostock
Printing and Binding: Malloy, Inc.
Cover Printing: John Pow Company

Library of Congress Cataloging-in-Publication Data
Howland, James S.
  Basic real analysis / James S. Howland.
      p. cm.
  Includes index.
  ISBN-13: 978-0-7637-7318-2 (hardcover)
  ISBN-10: 0-7637-7318-2 (ibid.)
  1. Mathematical analysis. 2. Numbers, Real. I. Title.
  QA300.H695 2010
  515—dc22
                                    2009027415

6048
Printed in the United States of America
13 12 11 10 09   10 9 8 7 6 5 4 3 2 1

For
Jimmy, Thomas, and Sarah

Contents

Preface xiii

Chapter 1 **Numbers** 1
    1.1 Real Analysis 1
    1.2 The Integers and Mathematical Induction 2
    1.3 The Real Numbers 5
    1.4 The Axiom of Continuity 8
    1.5 Supremum and Infimum 9
        1.5.1 Infinity 13
    1.6 The Archimedean Property 15
        1.6.1 *Irrational Numbers 17
    1.7 Supplementary Problems 18

Chapter 2 **Limits** 21
    2.1 Sequences 21
    2.2 The Limit of a Sequence 23
    2.3 Properties of Limits 26
    2.4 Infinite Limits 29
    2.5 The Monotone Sequence Theorem 31
        2.5.1 *The Number e 33
    2.6 The Bolzano–Weierstrass Theorem 35
    2.7 Cauchy Sequences 39
    2.8 *Application to Infinite Series 41
    2.9 *Limits Superior and Inferior 42
    2.10 Supplementary Problems 47

Chapter 3 **Continuity** 49
    3.1 Limits of Functions 49
    3.2 Limits and Sequences 52
    3.3 Continuity 55
    3.4 Infinite Limits 57
    3.5 *One-Sided Limits and Monotone Functions 59
    3.6 The Intermediate Value Theorem 62
    3.7 The Extreme Value Theorem 64
    3.8 Supplementary Problems 67

Chapter 4 Derivatives 69
- 4.1 The Derivative 69
- 4.2 Rules for Derivatives 72
- 4.3 The Critical Point Theorem 75
- 4.4 The Mean Value Theorem 76
- 4.5 *L'Hospital's Rule 81
- 4.6 Supplementary Problems 84

Chapter 5 Integrals 87
- 5.1 The Riemann Integral 87
- 5.2 Properties of the Integral 91
- 5.3 Riemann's Integrability Condition 94
- 5.4 Integrability Theorems 96
- 5.5 Uniform Continuity 99
- 5.6 Integrability of Continuous Functions 102
- 5.7 *Riemann Sums 103
- 5.8 The Fundamental Theorem 106
- 5.9 Substitution and Integration by Parts 108
- 5.10 *Improper Integrals 112
- 5.11 Supplementary Problems 116

Chapter 6 Infinite Series 123
- 6.1 Convergence 123
- 6.2 Series of Positive Terms 129
- 6.3 The Ratio and Root Tests 134
- 6.4 Absolute and Conditional Convergence 137
- 6.5 *Rearrangement of Series 142

Chapter 7 Uniform Convergence 147
- 7.1 Limits of Sequences of Functions 147
- 7.2 Uniform Convergence 149
- 7.3 Continuity 151
- 7.4 The Weierstrass M-Test 153
- 7.5 Integration 155
- 7.6 Differentiation 157
- 7.7 *Iterated Limits 158
- 7.8 Supplementary Problems 161

Chapter 8 Power Series 163
- 8.1 Power Series 163
  - 8.1.1 *Hadamard's Formula 165
- 8.2 Operations on Power Series 168
- 8.3 Taylor's Theorem 170
- 8.4 Supplementary Problems 177

Chapter 9   ***Further Topics in Series**   **179**
       9.1   *Summation by Parts   179
       9.2   *The Theorems of Abel and Tauber   181
       9.3   *The Integral Form of Taylor's Theorem   184
       9.4   *Kummer's Test   186

Appendix A   **Logic, Sets, and Functions**   **191**
       A.1   Logic   191
       A.2   Sets   191
       A.3   Functions   193
       A.4   Countable and Uncountable Sets   194

Appendix B   **The Topology of $\mathbb{R}$**   **199**
       B.1   Introduction   199
       B.2   Open Sets   199
            B.2.1   Continuity and Open Sets   201
       B.3   Closed Sets   202
            B.3.1   The Closure of a Set   203
       B.4   Compact Sets   207
            B.4.1   The Heine–Borel Theorem   208
            B.4.2   Compactness and Continuity   209

Appendix C   **Recommended Reading**   **211**
       C.1   Introduction   211

Index   **215**

Preface

Basic Real Analysis originates from notes for a course at the University of Virginia. The one-semester course is intended for students who have just finished Calculus. Its aim is to prove the basic theorems of Single Variable Calculus in as simple and direct a manner as possible. Any material I find to be of secondary importance is excluded. The goal is to get as far as uniform convergence, and I find this to be practical.

For use as a text, the material has been expanded somewhat to include definitions and topics that might reasonably be desired in a text of this level. A substantial number of problems have been added, to bring their number close to 300. However, I hope that the brevity and directness of the approach has survived. I would like to think that the text could be of use to a serious student as a supplement to a standard Calculus course.

I have tried to keep the level of sophistication low at the beginning, increasing somewhat as the text progresses. In part, this is simply a natural consequence of the material presented.

The first chapter introduces the concept of real analysis, beginning with the properties of various number systems, including natural numbers (positive integers) and real numbers. I then move on to discuss limits, including sequences, properties, and infinite limits. Continuity is discussed next, defining limits of functions as a continuous variable.

Derivatives are defined in Chapter 4, including rules for derivatives and critical theorems. Chapter 5 introduces Integrals, including the Riemann Integral, properties, theorems, and the concept of uniform continuity. Chapter 6 moves on to cover Infinite Series, discussing the difference between sequence and series, and convergence tests. This leads in to Chapter 7 on Uniform Convergence. This chapter covers iterated limits, different conditions on a sequence which ensure the preservation of various properties in the limit, integration, and differentiation.

Chapter 8, Power Series, demonstrates an important example of a sequence of functions, which is illustrated with the discussion of Taylor's Theorem. I conclude the text with a further discussion of series with Abel's Summation by Parts. I also illustrate the differences between Abel and Tauber's Theorems.

In the course, no time is devoted to point set topology, as this topic takes a good deal of time to digest, and may come at the expense of things like integration and series. In my opinion, this doctrine is better received after some sophistication has developed—perhaps after the chapter on power series. However, opinions and tastes differ, and I have included an Appendix on *Topology of the Reals* for those who wish to discuss these matters. It may be taken up at whatever time the instructor prefers.

Likewise, the course contains no explicit prefatory material on sets, functions or logic. I don't believe that this causes any difficulty. However, I have included a brief Appendix that summarizes what is needed, which can be taken as the first chapter if desired.

It seems to me—and to some colleagues to whom I put the question—that once sequential limits are discussed, the most easily understood definition of the limit of a function is through sequences. My students do not agree. They find the ϵ, δ-definition more natural. So I have made that the basic definition, and included the other as the "sequential criterion."

I wish to extend my appreciation to Tim Anderson, Melissa Potter, Amy Rose, Andrea Defonzo, Nina Hnatov, and the others at Jones and Bartlett who participated in the production of this book for their patience and assistance. Thanks also to my wife, Hope, for help in proofreading. But most of all, to my colleague, Almut Burchard, who used a preliminary version of this text in her class, go my sincere thanks for her suggestions and encouragement.

As usual, a starred section may be skipped without loss of continuity. A star on a problem indicates my subjective judgement that it may be more difficult than most.

Chapter 1
Numbers

1.1 Real Analysis

What is Real Analysis about?

Briefly stated, the purpose of this book is to provide a rigorous derivation of the basic theorems of Calculus from the properties of the system of Real Numbers.

Historically, the development that we will give is not the way that Calculus came about. As invented by Newton and Leibnitz in the 17th century, Calculus made use of the obscure notion of infinitesimals, for which there was no adequate foundation. In a famous critique, the philosopher Berkeley referred to infinitesimals as "the ghosts of departed quantities." This lack of a foundation had practical consequences. For example, the mathematicians of the 18th century operated with infinite series without an adequate theory of convergence. Sometimes they were successful in getting the right answer—and sometimes not. Since there was no theory to which to appeal, there was no criterion for when certain manipulations were correct.

It was only in the 19th century—some 200 years after its invention—that an adequate theory of Calculus was discovered. This theory, due to the efforts of Cauchy, Bolzano, Dedekind, Weierstrass, and others, reduces the notions of Calculus—limits, continuity, derivative, integral, infinite series, and the like—to the basic properties of the Real Number system. It is this theory, which is now called Real Analysis or Real Variable theory, that we are going to study.

You might suspect from the fact that it took 200 years to develop, and escaped the notice of people like Newton and Euler, that such a theory might not be entirely obvious and could involve ideas of considerable subtlety. And you would be correct. There is no escaping the fact that *Real Variable theory is one of the most difficult subjects for a student to master*. Therefore, be prepared to devote considerable time to your study. In particular, *you must understand every step of every proof.* Only when you are able to do this can you hope to construct proofs of your own.

The ultimate goal of this course is for you to be able—either by constructing a proof, or by knowing that you can do so—to decide for yourself when some manipulation of the objects of analysis (integrals, series, etc.) is justified. This will enable you to use the objects of analysis with a new freedom and confidence, and to begin to read and understand the more advanced mathematical literature that uses these objects.

You are assumed to be familiar with the properties of the usual elementary functions of Calculus—polynomials, rational functions, roots, exponentials, logarithms, trigonometric functions, and their inverses. They will be used in examples, but not in the proofs of the major theorems. You may also wish to review the standard notations and terminology of sets and functions that are summarized in Appendix A.

We begin our study in this chapter by recalling the properties of the various number systems. First, we consider the *Natural Numbers*, or positive integers, whose characteristic property is expressed in the *Principle of Mathematical Induction*. We then move on to the system of *Real Numbers*, whose characteristic property for the purpose of Analysis is expressed in the *Continuity Axiom*.

1.2 The Integers and Mathematical Induction

We begin with the system $\mathbb{N} = \{1, 2, 3, \dots\}$ of Natural Numbers, or positive integers. The characteristic property of $\mathbb{N}$ is expressed in the Principle of Mathematical Induction. In simple terms, it says that if you start with the number 1 and add 1 to it enough times, you will eventually get to any given natural number.

In order to introduce this property, we consider a specific problem. If we compute the successive sums of the odd integers, we find that

$$1 = 1 = 1^2$$
$$1 + 3 = 4 = 2^2$$
$$1 + 3 + 5 = 9 = 3^2$$
$$1 + 3 + 5 + 7 = 16 = 4^2.$$

We may therefore conjecture that the sum of the first n odd integers is n^2; that is,

$$1 + 3 + \cdots + (2n - 1) = n^2$$

for *every* natural number n.

How can we prove this? We may use the

Principle of Mathematical Induction *Let $P(n)$ be a proposition about the natural numbers. Assume that*

(a) *$P(1)$ is true; and*

(b) *if $P(n)$ is true for some fixed integer n, then $P(n+1)$ is also true.*

Then $P(n)$ is true for all natural numbers.

Example To prove the formula, let $P(n)$ be the proposition

$$P(n): \quad 1 + 3 + \ldots + (2n-1) = n^2.$$

Then

(a) $P(1)$ is true because $1 = 1^2$.

(b) Assume that $P(n)$ is true for some particular integer n. Then

$$1 + 3 + \ldots + (2n+1) = 1 + 3 + \ldots + (2n-1) + (2n+1)$$
$$= n^2 + (2n+1) = (n+1)^2.$$

Thus $P(n+1)$ is also true. Therefore, by the Principle of Induction, $P(n)$ is true for all n. ∎

As a second application, we prove the following:

Theorem 1 (Well-Ordering Principle) *Every nonempty set S of natural numbers has a least element.*

Proof Let $P(n)$ be the statement

$$P(n): \text{Every natural number less than } n \text{ is not in } S.$$

If $1 \in S$, then 1 is the least element of S, since it is the smallest natural number, and we are through.

If $1 \notin S$, then $P(1)$ is true. There must exist a natural number N such that $P(N)$ is true but $P(N+1)$ is not true. For otherwise, the Principle of Induction would imply that $P(n)$ is true for all n, which would mean that S is empty. But then N is the smallest element of S. (Why?) ∎

Chapter 1 ■ Numbers

1.2.1 Problems

1. Prove that $n < 2^n$ if $n \geq 1$.

2. Prove that for $n \geq 1$,
$$\sum_{k=1}^{n} k = 1 + 2 + \cdots + n = \frac{n(n+1)}{2}.$$

3. Prove that for $n \geq 1$,
$$\sum_{k=1}^{n} k^3 = 1^3 + 2^3 + 3^3 + \ldots + n^3 = \left(\frac{n(n+1)}{2}\right)^2.$$

4. *(Bernoulli's inequality)* Prove that for $n \geq 1$,
$$(1+x)^n \geq 1 + nx$$
if $x > -1$.

5. Prove Bernoulli's inequality for $n < 0$.

6. Prove that if a line of unit length is given, then a line of length $\sqrt{n}$ can be constructed for each n.

7. *(Binomial Theorem)* Define the Binomial coefficients
$$\binom{n}{k} = \frac{n!}{(n-k)!k!} = \frac{n(n-1)\cdots(n-k+1)}{1 \cdot 2 \cdots k}$$
and $\binom{n}{k} = 0$ for $k < 0$.

   (a) Prove that
$$\binom{n}{k} + \binom{n}{k-1} = \binom{n+1}{k}.$$

   (b) Prove by induction that
$$(x+y)^n = \sum_{k=0}^{n} \binom{n}{k} x^k y^{n-k}.$$

   (c) Prove that
$$\binom{n}{k} \leq \frac{n^k}{k!}.$$

8. Prove that
$$\sum_{k=1}^{n} k^2 = \frac{n(n+1)(2n+1)}{6}.$$

1.3 The Real Numbers

We turn next to the system $\mathbb{R}$ of Real Numbers, which are the raw material from which Analysis is built. All of the theorems and results of Calculus are deduced from the properties of the Real Numbers, which we shall list here.

These properties or Axioms come in three groups, concerned respectively with

1. Algebra,
2. Order, and
3. Continuity.

The *Algebraic* and *Order* properties are familiar from high school Algebra, but the key property of the reals for the purpose of Analysis is the *Continuity Axiom*. It is only a slight exaggeration to say that all of Real Analysis is based on this Axiom.

Axioms of Algebra

Algebraically, the Real Numbers form what is known in mathematical jargon as a *Field*. The briefest description of a Field is that it is something that satisfies the "usual rules of high school Algebra." In order to be precise, however, we shall spell out exactly which rules we have in mind.

Definition 1 A *Field* F is a set of objects for which an *addition* $a + b$, and a *multiplication* ab are defined satisfying the following rules:

1. (**Commutative Laws**)
$$a + b = b + a$$
$$ab = ba$$

2. (**Associative Laws**)
$$(a + b) + c = a + (b + c)$$
$$(ab)c = a(bc)$$

3. (**Distributive Law**)
$$a(b + c) = ab + ac$$

4. (**Zero and One**) There are two distinct special elements of F, denoted by 0 and 1, such that
$$a \cdot 1 = a$$

and
$$a + 0 = a$$
for every $a \in F$.

5. (**Negatives and Inverses**) For every $a \in F$, there is an element $-a$ such that
$$a + (-a) = 0.$$

For every $a \neq 0$ in F, there is an element $a^{-1} = 1/a$, called the *inverse* of a, such that
$$aa^{-1} = 1.$$

The first axiom is the following:

Axiom 1 *The set $\mathbb{R}$ of Real Numbers is a Field.*

There are a lot of algebraic consequences to these rules, such as $a \cdot 0 = 0$, and we assume that the reader is familiar with them.

The Real Numbers are by no means the only Field. Two familiar examples are the *Rational Numbers* $\mathbb{Q}$ and the *Complex Numbers* $\mathbb{C}$.

The *Integers*
$$\mathbb{Z} = \{\ldots -2, -1, 0, 1, 2, \ldots\}$$
are *not* a Field, because, for example, the number 2 has no inverse *in the set of Integers*. The inverse of 2 is, of course, the number 1/2, but this number *is not an integer*, and so is not in $\mathbb{Z}$. It is, indeed, the desire to have inverses of integers that leads to the Rationals.

Axioms of Order

One thing that sets the Reals apart from other Fields is that the Reals are *Ordered*. We shall assume the following:

Axiom 2 *There is an order $<$ defined on the Real Numbers $\mathbb{R}$ such that*

1. If $a < b$ and $b < c$, then $a < c$.

2. For every two real numbers, exactly one of the following holds:
$$a < b \qquad a = b \qquad b < a.$$

3. If $a < b$, then
$$a + c < b + c$$
for any c.

4. If $a < b$ and $c > 0$, then
$$ac < bc.$$

Again, the reader is assumed to be familiar with the simple consequences of such an ordering.

Definition 2 A Field with such an ordering is called an *Ordered Field*.

The Rationals $\mathbb{Q}$ are also an Ordered Field. However, the Complex Numbers $\mathbb{C}$ are *not* an Ordered Field. That is to say, it is impossible to invent a way to order the Complex Numbers so that this Axiom will hold (see Problem 4).

Absolute Value

We assume that the reader is familiar with the *absolute value function* defined by

$$|x| = \begin{cases} x & \text{if } x \geq 0 \\ -x & \text{if } x < 0. \end{cases}$$

It has the following properties

$$|x + y| \leq |x| + |y|$$
$$|xy| = |x|\,|y|$$
$$|cx| = c\,|x| \quad \text{if } c > 0.$$

The first of these properties is called the *triangle inequality*. It also holds for vectors, which accounts for the name.

It is very important to note that $|x - y|$ is the *distance between the numbers* x and y.

1.3.1 Problems

1. Prove that in an ordered field,

    (a) $0 \cdot x = 0$, for every x.
    (b) $x > 0$ iff $-x < 0$.

(c) $x^2 > 0$ if $x \neq 0$.

(d) $1 > 0$.

2. Prove that there is no ordering of the complex numbers under which they become an ordered field.

    (*Hint*: If there were such an ordering, would i be greater or less than 0?)

3. Prove that if $a < b + \epsilon$ for *every* positive real number ϵ, then $a \leq b$.

4. Prove that
$$||x| - |y|| \leq |x - y|.$$

(*Hint*: $x = (x - y) + y$.)

1.4 The Axiom of Continuity

The third Axiom sets the Reals apart from all other Fields. It is referred to as the *Continuity* or *Completeness* of the Real Numbers. *All Analysis is based on this fundamental fact.* There are many seemingly different ways to state it, and we shall meet several in the course of our study.

Geometrically, the Real Numbers are visualized as a continuous line. The Principle of Completeness is a *geometrical* property of this line. *It says that there are no "holes" in the line.* We have chosen a precise statement of this fact, which is due to Dedekind.

Axiom 3 (Dedekind) *Let the Real Numbers be divided into two nonempty, disjoint sets A and B such that*
$$a < b$$
for every $a \in A$ and every $b \in B$. Then there exists a real number p such that
$$a \leq p \leq b$$
for every $a \in A$ and $b \in B$.

We say in this case that p is determined by the *cut* (A, B).

Note that *the field of Rationals $\mathbb{Q}$ does not have the property of Completeness*. For example, let A be the set consisting of all negative numbers together with all positive numbers whose square is less than 2, and B the set of all positive numbers whose square is greater than or equal to 2. As we shall show in Theorem 6, the desired number p is $\sqrt{2}$, which is *not rational,* so if we are restricted to rational numbers alone, there is no number p lying between A and B.

1.5 Supremum and Infimum

The Supremum

Let S be a nonempty set of Real Numbers. We shall now define a number known as the *least upper bound* of S, or, in Latin, the *supremum* of S.

The best way to describe the supremum of S is to say that it wants to be the greatest element of S. In fact, if S has a greatest element, then that element is the supremum. However, many sets do not have greatest elements. For example, the *closed* interval $[0, 1]$ has as its largest element the number 1, but the *open* interval $(0, 1)$ has no largest element. Nevertheless, the number 1 seems to want to be the largest element of $(0, 1)$, if only it were actually in the set.

How can we describe this relationship of the number 1 to the set $(0, 1)$ in precise terms? What we can say is that 1 is the *smallest number that is bigger than every number of the set* $(0, 1)$; in other words, it is the *least upper bound* of the set.

Definition 3 Let S be a set of real numbers. A number M is an upper bound of S iff

$$s \leq M$$

for every $s \in S$. The set S is bounded above iff *S has an upper bound.*

The number M_0 is a least upper bound of S if

(a) M_0 is an upper bound of S, and

(b) M_0 is less than or equal to every upper bound of S.

Theorem 2 (Uniqueness) *A set S can have at most one least upper bound.*

Proof Let M_0 and M_1 be least upper bounds of S. We must prove that $M_0 = M_1$. Since M_1 is an upper bound, and M_0 a least upper bound, we have, $M_0 \leq M_1$. Similarly, $M_1 \leq M_0$. Hence, $M_0 = M_1$. □

We may therefore speak of *the* least upper bound of a set. The least upper bound of S is also called the *supremum* of S, and we write

$$M_0 = \sup S.$$

If the set S has a greatest element $s_{\max}$, then

$$\sup S = s_{\max}.$$

But a set may have a supremum even if it has no largest element.

Example 1 Both the closed interval $[0, 1]$ and the open interval $(0, 1)$ have

$$\sup [0, 1] = \sup (0, 1) = 1.$$

However, the number 1 is an element of $[0, 1]$, but not of $(0, 1)$. ∎

Example 2 The set

$$S = \left\{ \frac{1}{2}, \frac{2}{3}, \frac{3}{4}, \dots \right\} = \left\{ \frac{n}{n+1} : n \in \mathbb{N} \right\}$$

has the supremum

$$\sup S = 1,$$

which again is not in the set S. ∎

The Continuity Axiom ensures that a set with an upper bound actually has a least upper bound.

Theorem 3 (Least Upper Bound Theorem) *If a set S is bounded above, then it has a least upper bound.*

Proof Let B be the set of all upper bounds of S, and A the set of all numbers that are *not* upper bounds of S. Then (A, B) is a Dedekind cut of the reals. For if a is in A and b is in B, then there is a point s in S with $a < s$. But then $b \geq s > a$, since b is an upper bound of S.

Let p be the point determined by the cut (A, B). We claim that p is an upper bound of S. For if it is not, then there is a number $s \in S$ such that $p < s$. But

$$a = \frac{p + s}{2}$$

satisfies

$$p < a < s.$$

Since $a < s$ it is not an upper bound of S, so $a \in A$. But this contradicts the fact that $a \leq p$ by definition of p.

By definition, $p \leq b$ for all upper bounds b, so p is the least upper bound of S. □

The Least Upper Bound Theorem is not true for the rationals, because, for example, the set of positive numbers whose square is less than 2 has no *rational* least upper bound. Its least upper bound should be $\sqrt{2}$ but that number is *not rational*; it is one of the "holes" in the rationals that are filled in by the Axiom of Continuity.

Dedekind's Axiom can be proved from the Least Upper Bound Theorem, so this theorem is an equivalent version of the property of Completeness. It is frequently taken as an axiom instead of Dedekind's Axiom. We have, however, preferred to start with Dedekind's Axiom since it is geometrically more intuitive.

The Infimum

The *infimum* or *greatest lower bound* of a set S wants to be the smallest element of S. It is the largest number that is less than or equal to every element of the set.

Definition 4 *Let S be a set of real numbers. A number L is a lower bound of S iff*

$$s \geq L$$

for every $s \in S$. The set S is bounded below iff it has a lower bound. The number m_0 is a greatest lower bound of S iff

(a) *m_0 is a lower bound of S, and*

(b) *m_0 is greater than or equal to every lower bound of S.*

Corollary 1 (Uniqueness) *A set S can have at most one greatest lower bound.*

Proof This follows as in the proof of Theorem 2 (see Problem 4). □

We may therefore speak of *the* greatest lower bound of a set. The greatest lower bound of S is also called the *infimum* of S, and we write

$$m_0 = \inf S.$$

If the set S has a least element $s_{\min}$, then

$$\sup S = s_{\min}.$$

But a set may have an infimum even if it has no least element.

Example 3

(a) The set $\mathbb{N}$ of natural numbers has

$$\inf \mathbb{N} = 1.$$

(b) The open interval $(0, 1)$ and the closed interval $[0, 1]$ both have
$$\inf (0, 1) = \inf [0, 1] = 0.$$

(c) The set
$$S = \left\{1, \frac{1}{2}, \frac{1}{3}, \frac{1}{4}, \ldots\right\} = \left\{\frac{1}{n} : n \in \mathbb{N}\right\}$$
has the infimum
$$\inf S = 0.$$
■

The supremum and infimum are closely related. In fact, we have

Lemma 1 *If we define the set* $-S$ *to be*
$$-S = \{x : -x \in S\}$$
then
$$\inf S = -\sup(-S).$$

Proof See Problem 2. □

From this and the Least Upper Bound Theorem, there follows:

Corollary 2 *If a set* S *is bounded below, then it has a greatest lower bound.*

Definition 5 *The set* S *is* bounded *iff it has both an upper and a lower bound, or, equivalently, iff for some* M
$$|s| \leq M$$
for every $s \in S$.

The following simple result is very useful.

Proposition A *Let* S *be nonempty and bounded above, and* $M = \sup S$. *Then for every* $\epsilon > 0$, *there exists an element* s *of* S *such that*
$$M - \epsilon < s \leq M.$$

Proof See Problem 6. □

1.5.1 Infinity

Some sets, of course, do not have upper or lower bounds. For example, the set of natural numbers $\mathbb{N} = \{1, 2, 3, ...\}$ has no upper bound, the set of their negatives $\{-1, -2, -3, ...\}$ has no lower bound, and the set of integers

$$\mathbb{Z} = \{... -2, -1, 0, 1, 2, ...\}$$

has neither an upper nor a lower bound.

We introduce the notation ∞ to represent this case.

Definition 6 Let S be a nonempty set. We say that

$$\sup S = \infty$$

iff S has no upper bound. We say that

$$\inf S = -\infty$$

iff S has no lower bound.

We *never* have $\sup S = -\infty$ or $\inf S = \infty$ for a nonempty set.

Example 4 The set $\mathbb{N}$ of natural numbers has $\sup \mathbb{N} = \infty$. The set $\mathbb{Q}$ of rationals has

$$\sup \mathbb{Q} = \infty \quad \text{and} \quad \inf \mathbb{Q} = -\infty.$$

The following result is similar to Proposition A. ∎

Proposition B *If* $\sup S = \infty$, *then for every* $N > 0$, *there exists an element* s *of* S *such that* $s > N$.

Proof See Problem 7. □

    WARNING The statement that

$$\sup S = \infty$$

is just a *shorthand way of saying that the set S has no upper bound*. It *does not say* that $\sup S$ is equal to the "number" ∞.

It is important to realize that

$$\infty \text{ is not a number!}$$

That is, it is not possible to add the elements ∞ and $-\infty$ to the system of real numbers and define addition and multiplication in such a way that the usual rules of Algebra (i.e., the Field Axioms) hold.

This means that, while a useful shorthand,

the symbol ∞ cannot be manipulated like a number.

If one could do this, then, for example, either

$$1 + \infty = \infty = 2 + \infty$$

or

$$1 \cdot \infty = \infty = 2 \cdot \infty$$

would imply that $1 = 2$ by cancellation of ∞.

In this regard, ∞ differs from the imaginary unit i with

$$i^2 = -1$$

which, when adjoined to the reals $\mathbb{R}$, gives a system $\mathbb{C}$, the complex numbers, which is again a Field (although not an *Ordered* Field).

1.5.2 Problems

1. Find the supremum and infimum of the following sets.

   (a) $\{x : 0 < x \text{ and } x^2 < 3\}$
   (b) $\{x \in \mathbb{Q} : 0 < x^2 < 2\}$
   (c) $\{\frac{1}{n} : n = 1, 2, 3, \ldots\}$
   (d) $\{x : x^3 + 1 > 0\}$
   (e) $\{\sin n : n = 1, 2, 3, \ldots\}$
   (f) $\{(-1)^{n+1} n : n = 1, 2, 3, \ldots\}$
   (g) $\{x > 0 : \frac{x+1}{x}\}$
   (h) $\{\arctan x : x \in \mathbb{R}\}$

2. Prove that for any set S,

   $$\inf S = -\sup(-S)$$

   and

   $$\sup S = -\inf(-S).$$

   Include the cases where one of these is $\pm\infty$.

3. If $S_0 \subset S$, then $\sup S_0 \leq \sup S$.

4. Prove Corollary 1.

5. Prove Corollary 2 directly from Dedekind's Axiom.

6. Prove Proposition A.

7. Prove Proposition B.

8. Prove that the set $\{2^n : n \geq 1\}$ is unbounded above.

1.6 The Archimedean Property

The *Archimedean Property* of the Reals states that there are no *infinitely large* real numbers.

Theorem 4 (Archimedean Property) *For every real number a there exists a natural number N such that $a < N$.*

Proof Suppose that there is a real number a with $a > n$ for *every* natural number n. Then the set $\mathbb{N}$ of natural numbers is bounded above by a. By the Least Upper Bound Theorem, $\mathbb{N}$ has a least upper bound M. But then $M - 1$ cannot be an upper bound of $\mathbb{N}$, so there is a natural number n such that

$$M - 1 < n,$$

which implies that

$$M < n + 1.$$

This contradicts the fact that M is an upper bound of $\mathbb{N}$. $\square$

The Archimedean Property also implies that there are no *infinitely small* positive real numbers.

Corollary 3 *For every positive real number a there exists a natural number n such that $0 < 1/n < a$.*

Proof By the Archimedean Property, we may choose n such that $1/a < n$. Multiply this inequality by the positive number a/n to obtain $0 < 1/n < a$. $\square$

Another consequence of the Archimedean Property is that there is a rational number between any two real numbers. This is often referred to by saying that the Rationals are *dense* in the Reals.

Theorem 5 *Let $a < b$. Then there is a rational number r such that $a < r < b$.*

Proof By the Archimedean Property, choose a natural number n such that

$$0 < \frac{1}{n} < b - a$$

and an integer N greater than b. The set of Integers

$$\left\{k : \frac{k}{n} < b\right\}$$

contains no integer greater that nN, and so has a largest element m. Therefore

$$\frac{m}{n} < b \leq \frac{m+1}{n}$$

and so

$$a = b - (b - a) < b - \frac{1}{n} \leq \frac{m+1}{n} - \frac{1}{n} = \frac{m}{n}.$$

Thus

$$r = \frac{m}{n}$$

lies between a and b. $\square$

The following is a consequence of the Archimedean Property alone.

Proposition C *Let (A, B) be a Dedekind cut. Then for any positive number ϵ there exists a number $a \in A$ and a number $b \in B$ such that $b - a < \epsilon$.*

Proof By the Archimedean Property, there exists an integer n such that $0 < 1/n < \epsilon$. Consider the numbers of the form k/n. The set of Integers k such that k/n is in A is bounded above, and therefore has a largest element $a = k/n \in A$. But then $b = (k+1)/n \in B$, and

$$b - a = \frac{1}{n} < \epsilon.$$

$\square$

1.6.1 *Irrational Numbers

Dedekind's Axiom leads to the existence of irrational numbers; for example, the square root of 2.

Theorem 6 *There exists a unique number $p > 0$ with $p^2 = 2$.*

Proof Let A be the set consisting of all negative numbers and all positive numbers whose square is less than 2:

$$A = \{x \in \mathbb{R} : x < 0 \text{ or } x^2 < 0\}.$$

Then B is the set of all *positive* numbers whose square is greater than or equal to 2:

$$B = \{x \in \mathbb{R} : x > 0 \text{ and } x^2 \geq 0\}.$$

Let p be determined by the cut (A, B). By Proposition C, choose $a \in A$ and $b \in B$ with $a > 0$, $b < 4$, and

$$b - a < \frac{\epsilon}{4}.$$

Then $a < p < b$ implies that $a^2 < p^2 < b^2$, so that both p^2 and 2 lie between a^2 and b^2. Hence,

$$\left|p^2 - 2\right| < b^2 - a^2 = (b-a)(b+a) < \frac{\epsilon}{4} \cdot 4 = \epsilon.$$

Since $\epsilon > 0$ is arbitrary, we have $p^2 = 2$.

Uniqueness follows by Algebra. Let $r > 0$ be possibly another square root of 2. Then

$$0 = p^2 - r^2 = (p-r)(p+r).$$

Canceling $p + r$, which is positive, we have $p - r = 0$, or $p = r$. □

Irrationality of $\sqrt{2}$

Since it is quite possible that you have never seen a proof of the often-mentioned fact that $\sqrt{2}$ is irrational, we include one here.

Theorem 7 *There is no rational number whose square is 2.*

Proof The proof is based on the simple fact that the square of an odd number is odd. Suppose that
$$\sqrt{2} = \frac{p}{q}$$
for some integers. We will show that this leads to a contradiction.

Take the fraction in lowest terms, so that p and q *cannot both be even*. Then
$$p^2 = 2q^2$$
so that p must be even. However, writing $p = 2m$ leads to
$$2m^2 = q^2$$
so that q is also even. This is a contradiction. □

A second proof can be based on the unique factorization of integers into primes. Since the square of a number has an even number of prime factors, the equation
$$p^2 = 2q^2$$
is contradictory because the left side has an even number of prime factors, while the right side has an odd number.

1.6.2 Problems

1. Prove directly that $\mathbb{Q}$ is Archimedean.

2. Prove that between every two real numbers there is an *irrational* number. Use that $\sqrt{2}$ is irrational.

3. Let n be a positive integer and $c > 0$. Prove that there exists a unique number $r > 0$ such that $r^n = c$. We will write as usual $r = \sqrt[n]{c}$.

1.7 Supplementary Problems

1. Prove the Principle of Mathematical Induction from the Well-Ordering Principle.

2. Prove the *Principle of Complete Induction*:

    Let $P(n)$ be a statement about the natural number n. Suppose that

    (a) $P(1)$ is true, and

    (b) if $P(k)$ is true for all $k < n$, then $P(n)$ is true.

Then $P(n)$ is true for all natural numbers n.

(*Hint*: Consider the statement $S(n) =$ "$P(k)$ is true for all $k < n$.")

3. Prove that $1 + 1 + \cdots + 1$ is never 0 in an ordered field.

4. Let F be the set of all numbers of the form $a + b\sqrt{2}$ where a and b are rational numbers. Prove that F is a Field.

*5. Let Z_2 be the set consisting of the two elements 0, 1. Define multiplication and addition by the rules

$$x + 0 = x \qquad x \cdot 0 = 0$$
$$1 \cdot 1 = 1 \qquad 1 + 1 = 0.$$

Prove that Z_2 is a Field, but cannot be made into an Ordered Field.

6. Prove Dedekind's Axiom from the Least Upper Bound Theorem.

7. Let S and T be any sets. Prove that

(a) $\sup(-S) = -\inf S$, and

(b) $\sup[S + T] = \sup S + \sup T$, where

$$S + T = \{s + t : s \in S \text{ and } t \in T\}.$$

8. Let $f(x)$ and $g(x)$ be functions defined on S. Prove that

$$\sup\{f(x) + g(x) : x \in S\} \leq \sup\{f(x) : x \in S\} + \sup\{g(x) : x \in S\}.$$

Show by example that equality need not hold.

Chapter 2

Limits

2.1 Sequences

A *sequence*

$$a_1, a_2, \ldots, a_n, \ldots$$

is an infinite list of objects. It assigns to each positive integer n an object a_n, the nth item of the list, commonly called the nth *term*. More precisely, therefore, we may say that a *sequence* is a *function defined on the positive integers*.

The objects of the sequence may be anything. One may have sequences of numbers, sets, vectors, functions, or anything else. However, for the most part, we shall be concerned with sequences of *real numbers*.

One way to define a sequence is to give a *formula for the nth term*.

Example 1

(a) The sequence $1, 2, 3, 4, \ldots$ of positive integers may be defined by the formula

$$a_n = n.$$

(b) The sequence $2, 4, 6, 8, \ldots$ of even integers may be defined by the formula

$$a_n = 2n.$$

(c) The sequence $2, 4, 8, 16, \ldots$ of powers of 2 may be defined by the formula

$$a_n = 2^n.$$ ∎

A sequence may also be *defined inductively* by

(a) defining the first term a_1 (or the first few terms), and

(b) giving a method for computing the nth term from the terms preceding it.

Example 2

(a) The sequence $2, 4, 6, 8, \ldots$ of even integers may be defined inductively by
$$a_1 = 2 \qquad a_n = a_{n-1} + 2.$$

(b) The sequence $2, 4, 8, 16, \ldots$ of powers of 2 may be defined inductively by
$$a_1 = 2 \qquad a_n = 2\,a_{n-1}.$$

(c) The *Fibonacci numbers* f_n
$$1, 1, 2, 3, 5, 8, 13, \ldots$$
are defined inductively by
$$f_1 = f_2 = 1 \qquad f_{n+2} = f_{n+1} + f_n.$$ ∎

2.1.1 Problems

1. Give an inductive definition and formula for the sequence
$$1, 3, 5, 7, \ldots$$
of odd integers.

2. A sequence is defined by
$$a_0 = 2$$
$$a_1 = 1$$
$$a_{n+2} = a_{n+1} + 2a_n.$$

  (a) Compute the first five terms of the sequence.

  (b) Prove that
$$a_n = 2^n + (-1)^n.$$

3. Let a_n be defined recursively by
$$a_1 = 1$$
$$a_{n+1} = a_1 + a_2 + \cdots + a_n.$$
Find and prove a formula for a_n.

2.2 The Limit of a Sequence

We now need to discuss the *limit of a sequence*. That is, we want to define the meaning of the statement

$$\lim a_n = L.$$

Intuitively, a sequence a_n is said to *converge* to a number L if a_n gets close to L as n gets large. When this is the case, we say that L is the *limit of the sequence* a_n.

This description, however, will hardly pass muster as a precise definition of limit. We need to consider more closely exactly what is meant by what we have just said. To say that $\lim a_n = L$ means roughly that

"a_n is close to L if n is large."

But this is not precise enough for our purposes. The statement must be refined. Consider first the question: *"How close is a_n to L?"* The answer is that a_n gets closer and closer to L the larger that n gets. In fact,

"a_n will become as close to L as we wish if n is large."

That is, no matter how small a number ϵ we choose, say $\epsilon = 1/1000$ or $\epsilon = 1/1,000,000$, a_n will be closer to L than this number ϵ if n is large. Thus,

"For any number $\epsilon > 0$, we will have $|a_n - L| < \epsilon$ if n is large."

How large does n have to be? It must be *large enough*. There must eventually be some point after which $|a_n - L|$ is always less than ϵ. Also, it is not enough that $|a_n - L|$ be less than ϵ for *some* very large n. Rather, $|a_n - L|$ must be less than ϵ for *all sufficiently large* values of n; that is, for all values of n larger than some large integer N.

We must, therefore, have the following:

"For any number $\epsilon > 0$, there must exist a positive integer N such that $|a_n - L| < \epsilon$ if $n > N$."

Note that the smaller the number ϵ is taken, the larger the integer N will have to be.

This is the precise definition of the limit of a sequence.

Definition 1 The sequence a_n is said to converge to the number L iff for every positive number ϵ, there is a positive integer N such that $|a_n - L| < \epsilon$ whenever $n \geq N$.

In this case, we say that L is the *limit of the sequence* a_n, and write
$$L = \lim a_n.$$
We say that the *limit of a_n exists* if there is a number L such that $L = \lim a_n$. If the limit of a sequence exists, the sequence is said to be *convergent;* otherwise, it is *divergent*.

We may sometimes also use the notations
$$L = \lim_{n \to \infty} a_n \quad \text{or} \quad a_n \to L$$
instead of $L = \lim a_n$.

We apply this definition to some explicit examples.

Example 1 We will prove that
$$\lim \frac{1}{n} = 0.$$

Proof Let $\epsilon > 0$. By the Archimedean Property, we can choose a positive integer N such that
$$N > \frac{1}{\epsilon}.$$
If $n > N$, then
$$\left| \frac{1}{n} - 0 \right| = \frac{1}{n} < \frac{1}{N} < \epsilon.$$
By the definition, this proves that $\lim 1/n = 0$. ∎

Example 2 As a more complicated example, we will prove that
$$\lim \frac{n^2}{n^2 + n + 1} = 1.$$

Proof Let $\epsilon > 0$. We want to make
$$\left| \frac{n^2}{n^2 + n + 1} - 1 \right| = \left| \frac{-n - 1}{n^2 + n + 1} \right| = \frac{n + 1}{n^2 + n + 1}$$
less than ϵ.

This is rather more complicated than the previous example. What we want to do is to *estimate this expression*. That is, we will *replace it by something that is larger, but simpler, and easier to handle*. We do this in two steps. First we replace the *numerator* $n + 1$ by something *larger*, namely $2n$:
$$\frac{n + 1}{n^2 + n + 1} < \frac{2n}{n^2 + n + 1}.$$

Next, we replace the *denominator* n^2+n+1 by something *smaller*, namely n^2:

$$\frac{2n}{n^2+n+1} < \frac{2n}{n^2} = \frac{2}{n}.$$

Thus we get

$$\left|\frac{n^2}{n^2+n+1} - 1\right| < \frac{2}{n}.$$

Now choose N so that $N > 2/\epsilon$. Then, if $n \geq N$,

$$\left|\frac{n^2}{n^2+n+1} - 1\right| < \frac{2}{n} < \frac{2}{N} < \epsilon.$$
∎

Example 3 Finally, we prove that

$$\lim \frac{n^2}{n^2-1} = 1.$$

Proof Let $\epsilon > 0$. We have

$$\left|\frac{n^2}{n^2-1} - 1\right| = \left|\frac{n^2-n^2+1}{n^2-1}\right| = \frac{1}{n^2-1}.$$

We must again estimate the *denominator* from *below*. If $n \geq 2$, then $n^2 \geq 4$ and so $1 \leq n^2/4$. Hence,

$$n^2 - 1 \geq n^2 - \frac{n^2}{4} = \frac{3n^2}{4}$$

and

$$\left|\frac{n^2}{n^2-1} - 1\right| \leq \frac{1}{n^2-1} \leq \frac{4}{3n^2} < \frac{4}{n^2}.$$

Thus if we choose $N > 2/\sqrt{\epsilon}$, we get

$$\left|\frac{n^2}{n^2-1} - 1\right| < \frac{4}{n^2} < \frac{4}{(2/\sqrt{\epsilon})^2} = \epsilon.$$
∎

Important Remark about Estimation The method of estimation—that is, of *replacing a complicated expression by something that is larger, but simpler, and easier to handle*—is used over and over in Analysis.

2.2.1 Problems

1. If p is a positive integer, prove directly from the definition that

$$\lim \frac{1}{n^p} = 0.$$

2. Prove directly from the definition that

   (a) $\lim \dfrac{n^2}{n^2 + 1} = 1$,

   (b) $\lim \dfrac{n}{n^2 - 1} = 0$, and

   (c) $\lim \dfrac{n^2 + 1}{n^2 + n} = 1$.

3. Find the limit

$$\lim \left(\sqrt{n+1} - \sqrt{n} \right)$$

and prove that your answer is correct.

2.3 Properties of Limits

Armed with our precise definition, we are now able to give rigorous proofs of some simple, and familiar, properties of limits.

First, we show that there can be only one limit of a sequence.

Theorem 1 (Uniqueness) *The limit is unique if it exists.*

Proof We must prove that if $\lim a_n = L$ and $\lim a_n = M$, then $L = M$. Let $\epsilon > 0$. Choose N_1 such that

$$|a_n - L| < \frac{\epsilon}{2}$$

for $n \geq N_1$, and N_2 such that

$$|a_n - M| < \frac{\epsilon}{2}$$

for $n \geq N_2$. Then if $n > N = \max\{N_1, N_2\}$, we have

$$0 \leq |L - M| = |L - a_n + a_n - M| \leq |a_n - L| + |a_n - M| < \frac{\epsilon}{2} + \frac{\epsilon}{2} = \epsilon.$$

But a nonnegative number that is less than every positive number is zero. □

A sequence a_n is said to be *bounded* iff there is a number A such that
$$|a_n| \le A$$
for all n.

Theorem 2 *Every convergent sequence is bounded.*

Proof Let $\epsilon = 1$ and choose N such that
$$|a_n - L| < 1$$
for $n \ge N$. Then
$$|a_n| = |a_n - L + L| \le |a_n - L| + |L| < 1 + |L|$$
if $n \ge N$, while
$$|a_n| \le \max\{|a_1|, \ldots, |a_N|\}$$
if $1 \le n \le N$. Hence,
$$|a_n| \le A = \max\{|a_1|, \ldots, |a_N|, 1 + |L|\}$$
for all $n \ge 1$. $\square$

Theorem 3 (Basic Limit Theorem) *If $\lim a_n = L$ and $\lim b_n = M$, then*

(a) $\lim (a_n + b_n) = L + M$, *and*

(b) $\lim (a_n b_n) = LM$.

(c) *If in addition, $M \ne 0$, then*
$$\lim \frac{a_n}{b_n} = \frac{L}{M}.$$

Proof

(a) Let $\epsilon > 0$. Choose N_1 such that
$$|a_n - L| < \frac{\epsilon}{2}$$
when $n \ge N_1$, and choose N_2 such that
$$|b_n - M| < \frac{\epsilon}{2}$$
when $n \ge N_2$. If $N = \max(N_1, N_2)$, then when $n \ge N$, we have
$$|(a_n + b_n) - (L + M)| = |(a_n - L) + (b_n - M)|$$
$$\le |a_n - L| + |b_n - M| < \frac{\epsilon}{2} + \frac{\epsilon}{2} = \epsilon.$$

(b) By Theorem 2, there is an A such that
$$|a_n| \leq A.$$
Choose N_1 such that
$$|a_n - L| < \frac{\epsilon}{A + |M|}$$
for $n \geq N_1$ and N_2 such that
$$|b_n - M| < \frac{\epsilon}{A + |M|}$$
for $n \geq N_2$. Then if $n \geq N = \max(N_1, N_2)$, we have
$$\begin{aligned}|a_n b_n - LM| &= |a_n(b_n - M) + (a_n - L)M| \\ &\leq |a_n||b_n - M| + |a_n - L||M| \\ &< A\frac{\epsilon}{A + |M|} + \frac{\epsilon}{A + |M|}|M| = \epsilon.\end{aligned}$$

(c) See Problem 5. $\square$

2.3.1 Problems

1. Prove that if $a_n = a$ for all n, then $\lim a_n = a$.

2. If $\lim s_n = a$, prove that $\lim |s_n| = |a|$. Is the converse true?

3. Prove that

    (a) if $a_n \geq 0$ for all n, and $\lim a_n = a$, then $a \geq 0$, and

    (b) if $a_n \leq b_n$ for all n, $\lim a_n = a$ and $\lim b_n = b$, then $a \leq b$.
    (*Hint*: Use (a).)

4. (*Sandwich Theorem*) Assume that
$$\lim a_n = \lim b_n = c$$
and that $a_n \leq c_n \leq b_n$ for all n. Prove that
$$\lim c_n = c.$$
(Note that it is necessary to prove that *the limit exists* and that *it is equal to c*.)

5. Prove Theorem 3(c).

6. Prove that for every real number a there exists a sequence r_n of Rational Numbers such that $r_n \to a$.

7. For what values of x does
$$\lim \frac{x + x^n}{1 + x^n}$$
exist? Graph the limit as a function of x.

8. Prove directly from the definition that
$$\lim \frac{n^3 - n + 1}{n^3 + n + 1} = 1.$$

9. Prove directly from the definition that if $\lim a_n = a$, then
$$\lim a_n^2 = a^2.$$

10. Prove that if $\lim a_n = 0$ and b_n is bounded, then
$$\lim a_n b_n = 0.$$

2.4 Infinite Limits

Consider next the sequence
$$a_n = n.$$

The limit of this sequence does not exist. Its terms do not get closer and closer to any finite number. However, it *diverges in a special way*. Its terms grow larger and larger as n increases and eventually pass any finite bound whatever.

We describe this by saying that the sequence $a_n = n$ *tends to infinity*.

The precise definition is the following:

Definition 2 We say that
$$\lim a_n = \infty$$
if for every M there exists a positive integer N such that
$$a_n > M$$
for every $n \geq N$.

We say that
$$\lim a_n = -\infty$$
if for every M there exists a positive integer N such that
$$a_n < M$$
for every $n \geq N$.

Example Let us prove that
$$\lim (n^2 + 1) = \infty.$$

Proof Let $M > 0$, and choose $N > \sqrt{M}$. Then for $n > N$, we have
$$n^2 + 1 > n^2 > \left(\sqrt{M}\right)^2 = M. \qquad \blacksquare$$

WARNING Remember that *the symbol ∞ is not a number and cannot be manipulated like one*. A sequence that tends to ∞ or to $-\infty$ is a *divergent sequence. Tending to infinity is just a special way of being divergent.* It does not mean that the number a_n is close to the number infinity, since there is *no such number* as infinity. In particular, an expression like
$$|a_n - \infty| < \epsilon$$
is *absolute nonsense*. The author has unfortunately seen this expression on examination papers; but now that you have been warned, you will certainly never write such a thing!

2.4.1 Problems

1. Prove that if p is a positive integer, then $\lim n^p = \infty$.

2. Prove directly from the definition that
$$\lim \frac{n^2}{n+1} = \infty.$$

3. Prove that if $\lim a_n = \infty$ and $\lim b_n = b$ is finite, then
$$\lim a_n + b_n = \infty.$$

4. Prove that if $\lim a_n = \infty$ and $a_n \leq b_n$ for all n, then $\lim b_n = \infty$.

5. Prove that if $a_n \leq b_n$, $\lim a_n = a$ and $\lim b_n = b$, then $a \leq b$, in the case that a or b is ∞.

6. Prove that if $a_n > 0$, then $\lim a_n = 0$ iff
$$\lim \frac{1}{a_n} = \infty.$$

7. Use Bernoulli's inequality to prove that

   (a) $\lim a^n = \infty$ if $a > 1$. (*Hint*: Write $a = 1 + x$ with $x > 0$.)
   (b) $\lim a^n = 0$ if $|a| < 1$. (*Hint*: Use Problem 6.)

8. Prove that if a_n is increasing and unbounded above, then $\lim a_n = \infty$.

9. Does
$$\lim (-1)^n n = \infty?$$

2.5 The Monotone Sequence Theorem

A sequence a_n of real numbers is called *increasing* iff, for all n,
$$a_n \leq a_{n+1}.$$
It is called *decreasing* iff
$$a_n \geq a_{n+1}.$$
A sequence that is either increasing or decreasing is called a *monotone* sequence.
A sequence a_n is called *strictly increasing* iff, for all n,
$$a_n < a_{n+1}$$
and *strictly decreasing* iff
$$a_n > a_{n+1}.$$

Examples

(a) The sequences
$$1, 2, 3, 4, \ldots$$
and
$$1, \frac{1}{2}, \frac{1}{3}, \frac{1}{3}, \frac{1}{5}, \ldots$$
are respectively strictly increasing and strictly decreasing.

(b) The sequence
$$1, 1, 2, 2, 3, 3, \ldots$$
is increasing (or nondecreasing), but not strictly increasing.

(c) The sequence
$$1, -2, 3, -4, \ldots$$
given by $a_n = (-1)^{n+1} n$ is not monotone. ∎

Remark Some writers use the term *nondecreasing* for what we have called *increasing* and *nonincreasing* for what we have called *decreasing*. They then use *increasing* to mean what we have called *strictly increasing*. In this text, **we will not use the terms** *nondecreasing* and *nonincreasing*. In general, the reader is advised to ascertain in which sense the terms *increasing* and *decreasing* are used whenever they are encountered.

The following theorem is a third version of Completeness.

Theorem 4 (Monotone Sequence Theorem) *A bounded monotone sequence is convergent.*

Proof Consider the case that a_n is increasing. Let
$$L = \sup \{a_n : n \geq 1\}.$$
Let $\epsilon > 0$, and choose an N such that
$$L - \epsilon < a_N \leq L.$$
If $n \geq N$, then $a_n \geq a_N$, and so
$$|L - a_n| = L - a_n \leq L - a_N < \epsilon. \qquad \square$$

Cantor's Principle

A sequence $I_n = [a_n, b_n], \ldots$ of closed intervals is said to be *nested* iff
$$I_1 \supset I_2 \supset \ldots \supset I_n \supset \ldots$$
and the length $|I_n| = (b_n - a_n)$ of I_n tends to zero:
$$\lim |I_n| = \lim (b_n - a_n) = 0.$$

An important consequence of the Monotone Sequence Theorem is the following:

Theorem 5 (Cantor's Principle) *Every nested sequence of closed intervals contains a unique point p in its intersection.*

2.5 The Monotone Sequence Theorem

Proof If p and q are points in the intersection of the I_n, then since
$$|p - q| \leq (b_k - a_k) \to 0$$
we must have $p = q$, and there can be at most one point in the intersection.

To prove that such a point exists, note that the left and right endpoints of $I_n = [a_n, b_n]$ form respectively an increasing sequence a_n and a decreasing sequence b_n. By the Monotone Sequence Theorem, $\lim a_k = a$ and $\lim b_n = b$ exist. Since
$$a_n \leq a \leq b \leq b_n,$$
a and b are in I_n for every n (and, of course, $a = b$). $\square$

Cantor's Principle is yet another version—our fourth—of the Completeness Property.

2.5.1 *The Number e

Corollary 1 *The limit*
$$e = \lim \left(1 + \frac{1}{n}\right)^n$$
exists.

Proof Let
$$e_n = \left(1 + \frac{1}{n}\right)^n = \left(\frac{n+1}{n}\right)^n.$$

We have
$$\frac{e_n}{e_{n-1}} = \left(\frac{n+1}{n}\right)^n \left(\frac{n}{n-1}\right)^{-n} \left(\frac{n}{n-1}\right)$$
$$= \left(\frac{n+1}{n} \cdot \frac{n-1}{n}\right)^n \left(\frac{n}{n-1}\right)$$
$$= \left(1 - \frac{1}{n^2}\right)^n \left(\frac{n}{n-1}\right)$$
$$> \left(1 - \frac{n}{n^2}\right)\left(\frac{n}{n-1}\right) = 1$$

where we have used Bernoulli's inequality at the last step. Thus $e_{n-1} < e_n$, so that e_n is increasing, and $e_n > e_1 = 2$.

To show that the sequence e_n is bounded, observe that by the Binomial Theorem and Part (c) of Problem 7 of Section 1.2,

$$e_n = \left(1 + \frac{1}{n}\right)^n = \sum_{k=0}^{n} \binom{n}{k} \frac{1}{n^k} \leq \sum_{k=0}^{n} \frac{1}{k!}$$

$$= 1 + 1 + \frac{1}{2!} + \frac{1}{3!} + \cdots + \frac{1}{k!}$$

$$< 1 + 1 + \frac{1}{2} + \frac{1}{2^2} + \cdots + \frac{1}{2^k} < 3.$$

Therefore, $2 < e_n < 3$ is an increasing sequence. It follows that the limit

$$e = \lim \left(1 + \frac{1}{n}\right)^n$$

exists and is a number between 2 and 3. Actually, of course, $e = 2.71828\ldots$ to five places. We will see how to compute e in Chapter 8. □

2.5.2 Problems

1. Which of the following sequences are increasing? Strictly increasing? Which are not monotone?

   (a) $1, 1, 1, 1, \ldots$
   (b) $-1, -2, -3, -4, \ldots$
   (c) $a_n = n^2$
   (d) $a_n = (-1)^{n+1} \frac{1}{n}$
   (e) $a_n = n^2 + (-1)^n n$

2. Prove that if $M = \sup S$ for some nonempty bounded set S, then there exists an *increasing* sequence s_n of points of S such that

   $$\lim s_n = M.$$

3. If the set S is unbounded above (i.e., $\sup S = \infty$), then there exists an *increasing* sequence s_n of points of S such that

   $$\lim s_n = \infty.$$

4. Prove that if $a > 0$, then $\lim a^{1/n} = 1$.

5. Prove that the sequence $1, \sqrt{2}, \sqrt{2\sqrt{2}}, \sqrt{2\sqrt{2\sqrt{2}}}, \ldots$ defined by

   $$a_0 = 1$$
   $$a_{n+1} = \sqrt{2a_n}$$

   has a limit. Find its value.

6. Prove that the sequence
$$a_n = \frac{1}{n+1} + \frac{1}{n+2} + \cdots + \frac{1}{2n}$$
converges to a limit a with $0 \leq a \leq 1$. (Actually, $a = \log 2$. See Problem 5 of Section 5.7.)

7. Prove that the following sequences converge:
   (a) $a_n = \dfrac{n^n}{2}$
   (b) $b_n = \dfrac{1 \cdot 3 \cdot 5 \cdots (2n-1)}{2 \cdot 4 \cdot 6 \cdots (2n)}.$

2.6 The Bolzano–Weierstrass Theorem

Subsequences

We now introduce the simple, but important notion of *subsequence*. A subsequence of a sequence a_n is a sequence obtained from a_n by omitting some of its terms.

Example 1

(a) The sequence
$$2, 4, 6, 8, \ldots$$
of positive even integers is obtained from the sequence
$$1, 2, 3, 4, \ldots$$
of positive integers by omitting every other term, and is therefore a subsequence of it.

(b) The sequence
$$1, \frac{1}{2}, \frac{1}{4}, \frac{1}{8}, \ldots, \frac{1}{2^n}, \ldots$$
is a subsequence of the sequence
$$1, \frac{1}{2}, \frac{1}{3}, \frac{1}{4}, \ldots, \frac{1}{n}, \ldots$$
∎

A subsequence b_k of a_n may always be written as $b_k = a_{n_k}$ where n_k is a *strictly increasing sequence of positive integers.*

Example 2

(a) In Example 1(a), if
$$a_n = n$$
is the sequence of positive integers, then the sequence of positive even integers is
$$b_k = 2k$$
so that
$$n_k = 2k.$$

(b) In Example 1(b),
$$a_n = \frac{1}{n}$$
and
$$n_k = 2^k.$$ ∎

Theorem 6 *If a_n converges to L, then every subsequence of a_n converges to L.*

Proof Let a_{n_k} be a subsequence of a_n. Let $\epsilon > 0$, and choose N such that $|a_n - L| < \epsilon$ whenever $n > N$. Choose K so that $n_K > N$. Then for $k > K$, we have $n_k > n_K > N$ and hence
$$|a_{n_k} - L| < \epsilon.$$ □

A subsequence of a_n may converge when a_n does not. For example, the sequence
$$a_n = (-1)^{n+1};$$
that is,
$$1, -1, 1, -1, 1, \ldots$$
does not converge, but its subsequence
$$a_{2n-1} = (-1)^{2n} = 1$$
converges to the limit $L = 1$.

Corollary 2 *If $0 < a < 1$, then*
$$\lim a^n = 0.$$

Proof We have
$$0 < a^{n+1} = a \cdot a^n < a^n$$
so the sequence a^n is decreasing and bounded below. By the Monotone Sequence Theorem, there is a number L such that $a^n \to L$. But
$$a^{2n} = a^n a^n. \tag{1}$$

Since the subsequence a^{2n} also converges to a by the preceding theorem, we obtain by passing to the limit in (1)
$$L = L^2.$$

The only numbers satisfying this equation are $L = 0$ and $L = 1$. Since the limit is clearly less than a, it cannot be 1, so
$$\lim a^n = L = 0. \qquad \square$$

The Bolzano–Weierstrass Theorem

The Bolzano–Weierstrass Theorem is one of the most important theorems of Real Analysis. It is one version of an important property called *Compactness*, and is behind almost all of the most difficult theorems that we shall prove.

Theorem 7 (Bolzano–Weierstrass) *Every bounded sequence has a convergent subsequence.*

Proof The proof of this theorem uses a device known as the *Method of Bisection*.

Let c_n be a bounded sequence, and choose C such that
$$|c_n| \leq C$$
so that the sequence c_n is entirely contained in the interval $I_0 = [-C, C]$. Divide the interval I_0 into two equal halves. At least one of these halves must contain an infinite number of terms of c_n. Choose I_1 to be such a half-sized interval. The length of I_1 is C; that is, half the length of I_0. Choose a point c_{n_1} in I_1. This can be done because I_1 contains an infinite number of terms of the sequence.

Continue this process. Bisect I_1, and choose I_2 to be one of the half intervals of I_1 containing an infinite number of terms of c_n. Choose $n_2 > n_1$ such that c_{n_2} is in I_2, which is again possible because I_2 contains an infinite number of terms of the sequence.

Continuing in this fashion, we obtain for every index k an interval $I_k = [a_k, b_k]$ of length

$$b_k - a_k = \frac{2C}{2^k} = \frac{C}{2^{k-1}}$$

and a point

$$c_{n_k} \in I_k$$

with $n_k < n_{k+1}$ and $I_k \supset I_{k+1}$.

By Cantor's Principle, there exists a point c in the intersection of the I_k's. Since both c and c_{n_k} are in I_k, we have

$$|c_{n_k} - c| \leq b_k - a_k$$

so that

$$\lim c_{n_k} = c. \qquad \square$$

The Bolzano–Weierstrass Theorem is a fifth version of Completeness.

2.6.1 Problems

1. Some of the following sequences are subsequences of others. Which are so related?

    (a) $1, -1, 1, -1, \ldots$
    (b) $-1, 1, -1, 1, \ldots$
    (c) $1, \frac{1}{2}, \frac{1}{3}, \frac{1}{4}, \ldots$
    (d) $1, \frac{1}{2}, \frac{1}{4}, \frac{1}{9}, \frac{1}{16}, \ldots$
    (e) $1, 0, \frac{1}{2}, 0, \frac{1}{3}, 0, \frac{1}{4}, \ldots$

2. Prove that $\lim a_n = L$ iff every subsequence a_{n_k} of a_n has a subsequence $a_{n_{k_j}}$ converging to L.

3. Prove that a *strictly* increasing sequence of *positive integers* tends to ∞.

*4. We have so far seen five versions of the Completeness Property of ordered fields:

Dedekind's Axiom,

The Least Upper Bound Theorem,

The Monotone Sequence Theorem,

Cantor's Principle, and the

Bolzano–Weierstrass Theorem.

Each has been proved from the one preceding it in this list.

In order to prove that they are all equivalent, all we need to do is to deduce Dedekind's Axiom from the Bolzano–Weierstrass Theorem. This may be done in two steps.

(a) Prove that the Bolzano–Weierstrass Theorem implies the Archimedean Property. Proposition C of Chapter 1 now follows, since it was proved from the Archimedean Property alone.

(b) Prove that the Bolzano–Weierstrass Theorem and Proposition C imply Dedekind's Axiom.

2.7 Cauchy Sequences

We shall introduce one more version, our sixth, of Completeness.

Definition 3 A sequence a_n is a *Cauchy sequence* iff for every $\epsilon > 0$, there exists a positive integer N such that

$$|a_n - a_m| < \epsilon$$

whenever $n, m \geq N$.

The idea here is that for large enough n, the terms of the sequence are all contained in an arbitrarily small interval.

Lemma 1 *Every Cauchy sequence is bounded.*

Proof Let $\epsilon = 1$, and choose N such that

$$|a_n - a_m| < 1$$

for $n, m \geq N$. Then for $n \geq N$

$$|a_n| = |a_n - a_N + a_N| \leq |a_N| + |a_n - a_N| \leq |a_N| + 1.$$

Hence,
$$|a_n| \leq \max\{|a_1|, \ldots, |a_{N-1}|, |a_N| + 1\}$$
for all n. □

Theorem 8 (Cauchy's Convergence Criterion) *A sequence is convergent iff it is a Cauchy sequence.*

Proof Assume that a_n is convergent, and let $L = \lim a_n$. Given $\epsilon > 0$, choose N such that
$$|a_n - L| < \frac{\epsilon}{2}$$
whenever $n \geq N$. Then
$$|a_n - a_m| = |a_n - L + L - a_m| \leq |a_n - L| + |a_m - L| < \frac{\epsilon}{2} + \frac{\epsilon}{2} = \epsilon$$
if $n, m \geq N$, so that a_n is a Cauchy sequence.

Conversely, suppose that a_n is a Cauchy sequence. By Lemma 1, a_n is bounded. Therefore, by the Bolzano–Weierstrass Theorem, a_n has a subsequence a_{n_k} that converges to some number L.

We claim that the full sequence a_n also converges to L. Let $\epsilon > 0$ and choose N so that
$$|a_n - a_m| < \frac{\epsilon}{2}$$
for $n, m \geq N$. If k is chosen so that $n_k \geq N$, then
$$|a_n - L| \leq |a_{n_k} - L| + |a_n - a_{n_k}| < \frac{\epsilon}{2} + \frac{\epsilon}{2} = \epsilon.$$
Hence, $\lim a_n = L$. □

2.7.1 Problems

*1. Prove that a bounded increasing sequence is a Cauchy sequence. Conclude that Cauchy's Convergence Criterion implies the Monotone Sequence Theorem.

*2. If $|a_{n+1} - a_n| \leq \theta |a_n - a_{n-1}|$, where $0 < \theta < 1$, then $\lim a_n$ exists.

3. Prove Cantor's Principle from Cauchy's Criterion by showing that the endpoints are Cauchy sequences.

Since we proved Cauchy's Theorem from Cantor's Principle, this establishes the equivalence of Cauchy's Criterion with Cantor's Principle, and hence with the other five versions of Completeness.

2.8　*Application to Infinite Series

We shall show next how to apply the Monotone Convergence Theorem and Cauchy's Criterion to the theory of infinite series. Our treatment is quite brief, since we only wish to show how these results are used, and not to develop the theory of series here.

This section may well be skipped for the present, since we cover the same subjects much more completely in Chapter 6.

In ordinary English, the terms *sequence* and *series* are practically synonymous. In Mathematics, however, they are entirely different. A *sequence* a_n is an *infinite list* of numbers (or other objects). An *infinite series*

$$\sum_{n=1}^{\infty} a_n = a_1 + a_2 + \cdots + a_n + \cdots$$

is an *infinite sum*, the sum of all the numbers of the sequence a_n.

To make sense of this, we simply *add up the first n terms, and take the limit as n tends to infinity*. More precisely, we define the *nth partial sum s_n* of the series to be the *sum of the first n terms*:

$$s_n = a_1 + a_2 + \cdots + a_n = \sum_{k=1}^{n} a_k.$$

The partial sums of a series form a sequence,

$$s_1, s_2, \ldots, s_n, \ldots$$

the limit of which is the sum of the series. The series $\sum_{n=1}^{\infty} a_n$ is *convergent* iff the limit

$$s = \lim s_n$$

of the sequence of partial sums exists. The number s is called the sum of the series, and we write

$$s = \sum_{n=1}^{\infty} a_n = a_1 + a_2 + \cdots + a_n + \cdots.$$

If a series is not convergent, it is called *divergent*. In particular, the series is divergent if $\lim s_n = \pm\infty$.

Series of Positive Terms

We say that $\sum_{n=1}^{\infty} a_n$ is a *series of positive terms* iff $a_n \geq 0$ for all n.

Theorem *A series of positive terms is convergent iff its sequence of partial sums is bounded.*

Proof Since we have
$$s_{n+1} = (a_1 + a_2 + \ldots + a_n) + a_{n+1} \geq a_1 + a_2 + \ldots + a_n = s_n$$
the sequence of partial sums is increasing. According to the Monotone Sequence Theorem, the increasing sequence s_n converges iff it is bounded. □

Absolute Convergence

The series $\sum_{n=1}^{\infty} a_n$ *converges absolutely* iff the series
$$\sum_{n=1}^{\infty} |a_n|$$
of absolute values converges.

Theorem *If a series converges absolutely, then it converges.*

Proof Let
$$\sigma_n = |a_1| + |a_2| + \cdots + |a_n|$$
be the partial sums of the series of absolute values. Since $\sum_{n=1}^{\infty} |a_n|$ is convergent, σ_n is a Cauchy sequence. Thus, for $m > n$,
$$|s_m - s_n| = \left| \sum_{k=n+1}^{m} a_k \right| \leq \sum_{k=n+1}^{m} |a_k| = \sigma_m - \sigma_n \to 0.$$
Hence, s_n is a Cauchy sequence as well. □

2.9 *Limits Superior and Inferior

In cases where the limit does not exist there are two partial replacements that are sometimes useful: the *limit superior* and the *limit inferior*. As we have seen, a given sequence a_n may have subsequences converging to many different limits, including $\pm\infty$.

In simple terms, *the limit superior of a_n is the largest possible limit to which a subsequence of a_n can converge.* Similarly, *the limit inferior of a_n is the smallest possible limit to which a subsequence of a_n can converge.*

The precise definition looks a little different, and goes like this. Fix n, and let b_n be the *supremum* of all values of a_k for $k \geq n$:

$$b_n = \sup_{k \geq n} a_k.$$

Since b_{n+1} is the supremum over a smaller set, we have $b_{n+1} \leq b_n$, so that b_n is a *decreasing* sequence. By the Monotone Sequence Theorem, it must have a (possibly infinite) limit. This limit is the limit superior:

$$\limsup a_n = \lim_{n \to \infty} \sup_{k \geq n} a_k.$$

Note that the limit superior may be infinity in the case that $b_n = \infty$ for every n, which is quite possible.

In a similar manner, the limit inferior is defined by taking the *infimum* over $k \geq n$ to get the *increasing* sequence

$$c_n = \inf_{k \geq n} a_k,$$

which must also have a (possibly infinite) limit. This limit is the *limit inferior*:

$$\liminf a_n = \lim_{n \to \infty} \inf_{k \geq n} a_k.$$

Either of these limits may be equal to ∞ or $-\infty$.

The following alternative notations are sometimes used:

$$\overline{\lim} a_n = \limsup a_n$$
$$\underline{\lim} a_n = \liminf a_n.$$

Theorem 9 *The limits superior and inferior exist for every sequence a_n, and satisfy*

$$\liminf a_n \leq \limsup a_n. \qquad (2)$$

Proof We have already proved existence above. We have, clearly,

$$c_n \leq a_n \leq b_n \qquad (3)$$

from which (2) follows. □

Theorem 10 *The (possibly infinite) limit of a_n exists iff*

$$\liminf a_n = \limsup a_n = L$$

in which case

$$\lim a_n = L.$$

Proof Suppose that
$$\liminf a_n = \limsup a_n;$$
that is, suppose that b_n and c_n converge to the same (finite or infinite) limit L. Then by (3), and the Sandwich Theorem, a_n must converge to L as well.

Conversely, suppose that the *finite* limit $\lim a_n = L$ exists. Then given $\epsilon > 0$, there exists N such that for $n \geq N$,
$$L - \epsilon < a_n < L + \epsilon.$$
Taking the sup over $k \geq n$ gives
$$L - \epsilon < b_n = \sup_{k \geq n} a_k \leq L + \epsilon,$$
which implies that,
$$|b_n - L| \leq \epsilon$$
for $n \geq N$. Thus,
$$\limsup a_n = \lim b_n = L.$$
Similarly for $\liminf a_n$.

The proof for $L = \infty$ is deferred to Problem 2. □

Theorem 11 *The limit superior is the largest number to which a subsequence of a_n converges.*

Proof Let $L = \limsup a_n$. We must show two things:

1. that a subsequence of a_n converges to L, and
2. that no subsequence of a_n can converge to a larger number.

For (1), we may construct a subsequence converging to L as follows. Let $\epsilon = 1/k > 0$. Since b_n converges to L, we may choose N such that
$$|b_n - L| < \frac{\epsilon}{2}$$
for $n > N$. Since
$$b_n = \sup_{k \geq n} a_k$$
we may then choose $n_k \geq n$ such that
$$|a_{n_k} - b_n| < \frac{\epsilon}{2}.$$

Then we have

$$|a_{n_k} - L| \leq |a_{n_k} - b_n| + |b_n - L| < \epsilon = \frac{1}{k}.$$

Letting $k \to \infty$ gives $\lim a_{n_k} = L$.

For (2), we simply note that for any subsequence a_{n_k}

$$a_{n_k} \leq b_{n_k}$$

so that

$$\lim a_{n_k} \leq \lim b_{n_k} = \limsup a_n. \qquad \square$$

Similarly, we have

Theorem 12 *The limit inferior is the smallest number to which a subsequence of a_n converges.*

Proof See Problem 3. $\qquad \square$

The Cesaro Limit

As a typical application of the limit superior, we discuss the Cesaro limit of a sequence. Let a_n be a sequence and

$$\bar{a}_n = \frac{a_1 + a_2 + \cdots + a_n}{n}$$

the arithmetic mean (i.e., the average) of its first n elements.

The limit $\lim \bar{a}_n$ of the sequence of arithmetic means is called the *Cesaro Limit* of the sequence a_n. It is equal to $\lim a_n$ when this limit exists, but may exist when $\lim a_n$ does not.

Theorem 13 *If $\lim a_n = a$, then $\lim \bar{a}_n = a$.*

Proof If $c_n = a_n - a$, then $\bar{c}_n = \bar{a}_n - a$, and $c_n \to 0$. We need to show that $\bar{c}_n \to 0$. Let $\epsilon > 0$, and choose N such that $|c_n| < \epsilon$ if $n > N$. Let $|c_n| \leq C$. Then

$$|\bar{c}_n| \leq \left| \frac{c_1 + c_2 + \cdots + c_N}{n} \right| + \left| \frac{c_N + c_{N+1} + \cdots + c_n}{n} \right|$$

$$\leq \frac{CN}{n} + \frac{n-N}{n}\epsilon.$$

Therefore,
$$0 \le \limsup |\bar{c}_n| \le \limsup \left(\frac{CN}{n} + \frac{n-N}{n}\epsilon \right) = \epsilon.$$

Since $\epsilon > 0$ is arbitrary, $\lim |\bar{c}_n| = 0$. □

Example The limit of the sequence a_n given by $1, 0, 1, 0, 1, \ldots$ does not exist. However, its arithmetic means are

$$\bar{a}_{2n} = \frac{1}{2} \quad \text{and} \quad \bar{a}_{2n+1} = \frac{1}{2} + \frac{1}{n}$$

so that the Cesaro limit is

$$\lim \bar{a}_n = \frac{1}{2}.$$ ■

2.9.1 Problems

1. Find the limits superior and inferior of the following sequences.

   (a) $(-1)^{n+1}$
   (b) $(-1)^{n+1} n$
   (c) $(-1)^{n+1} \frac{1}{n}$
   (d) $\sin(n)$

2. Prove Theorem 10 in the case where the limit is infinite.

3. Prove Theorem 12.

4. Show that if $\lim a_n = a > 0$, then
$$\limsup (a_n b_n) = a \limsup b_n.$$
   What happens if $a < 0$?

5. Prove that if $a_n > 0$ and $\lim a_n = a > 0$, then the sequence
$$g_n = (a_1 a_2 \cdots a_n)^{1/n}$$
   of geometric means converges to a.

2.10 Supplementary Problems

1. Show that the sequence of Fibonacci numbers is given by the formula
$$f_n = \frac{1}{\sqrt{5}}\left[\left(\frac{1+\sqrt{5}}{2}\right)^n - \left(\frac{1-\sqrt{5}}{2}\right)^n\right].$$

   (*Hint*: Verify that these numbers satisfy the inductive definition.)

2. Find the limit
$$\lim \left(\sqrt{n^2+1} - n\right)$$
   and prove that your answer is correct.

*3. (a) Prove that if $a_n > 0$ and
$$\lim \frac{a_{n+1}}{a_n} = L,$$
   then
$$\sqrt[n]{a_n} = L.$$

   (b) Find
$$\lim \sqrt[n]{\frac{n!}{n^n}}.$$

   (c) Show that the converse of (a) is false. (*Hint*: Try $a_n = e^{(-1)^n \sqrt{n}}$.)

4. Prove that if a_n does not have a convergent subsequence, then
$$\lim |a_n| = \infty.$$

5. Let S be a set of real numbers. Prove that if every sequence in S has a convergent subsequence, then S must be bounded.

6. Prove that there exists a *strictly* increasing sequence n_k of *positive integers* such that
$$\lim \sin n_k$$
   exists.

*7. Prove that every sequence has a monotone subsequence. Deduce the Bolzano–Weierstrass Theorem from this fact and the Monotone Sequence Theorem.

*8. Define a "generalized limit $L(a_n)$" to be the average value of the limit superior and the limit inferior:

$$L(a_n) = \frac{1}{2}\left[\limsup a_n + \liminf a_n\right].$$

Is this likely to be a useful definition? (*Hint*: What properties of the limit does it fail to have? Is it linear? Multiplicative?)

Chapter 3

Continuity

3.1 Limits of Functions

We shall now define the limit of a function of a continuous variable. The following is sometimes known as the ε, δ-*definition*.

Definition 1 Let a be a point of an interval I, and $f(x)$ a function defined on I, except possibly at the point a itself. We say that

$$\lim_{x \to a} f(x) = L$$

iff for every positive number ϵ, there exists a positive number δ such that

$$|f(x) - L| < \epsilon$$

whenever $0 < |x - a| < \delta$ and $x \in I$.

Remark Note that *the limit L does not depend in any way on the actual value of f at the point a*, which may not even be defined. Thus, we say that

$$\lim_{x \to 0} \frac{\sin x}{x} = 1$$

although $\sin x / x$ is not defined at $x = 0$.

Example We will prove that

$$\lim_{x \to 1} \frac{2x}{1 + x^2} = 1.$$

Proof Let $\epsilon > 0$, and choose $\delta = \sqrt{\epsilon}$. Then for $|x - 1| < \delta$, we have

$$\left| \frac{2x}{1+x^2} - 1 \right| = \left| \frac{2x - 1 - x^2}{1+x^2} \right| = \frac{|x-1|^2}{1+x^2} \leq |x-1|^2 < \delta^2 = \epsilon.$$ ∎

Theorem 1 (Uniqueness) *If the limit exists, it is unique.*

Proof Suppose that

$$\lim_{x \to a} f(x) = L \quad \text{and} \quad \lim_{x \to a} f(x) = M.$$

Let $\epsilon > 0$. Choose $\delta_1 > 0$ such that

$$|f(x) - L| < \epsilon$$

if $0 < |x - a| < \delta_1$, and choose $\delta_2 > 0$ such that

$$|f(x) - M| < \epsilon$$

if $0 < |x - a| < \delta_2$. Then, for $|x - a| < \min(\delta_1, \delta_2)$, we have

$$|L - M| < |f(x) - L| + |f(x) - M| < \frac{\epsilon}{2} + \frac{\epsilon}{2} = \epsilon.$$

Since ϵ is arbitrary, $L = M$. □

The proof of the next theorem is quite similar to its sequential version.

Theorem 2 (Basic Limit Theorem) *If*

$$\lim_{x \to a} f(x) = L \quad \text{and} \quad \lim_{x \to a} g(x) = M$$

exist, then

(a)

$$\lim_{x \to a} (f(x) + g(x)) = L + M$$

and

(b)

$$\lim_{x \to a} f(x)g(x) = LM$$

(c) *Moreover, if $M \neq 0$, then*

$$\lim_{x \to a} \frac{f(x)}{g(x)} = \frac{L}{M}.$$

Remark The conclusion of the theorem includes the statement that these limits exist.

Proof (a) Let $\epsilon > 0$. Choose $\delta_1 > 0$ such that

$$|f(x) - L| < \frac{\epsilon}{2}$$

if $0 < |x - a| < \delta_1$. Choose $\delta_2 > 0$ such that

$$|g(x) - M| < \frac{\epsilon}{2}$$

if $0 < |x - a| < \delta_2$. Let $\delta = \min(\delta_1, \delta_2)$. Then

$$|f(x) + g(x) - (L + M)| \leq |f(x) - L| + |g(x) - M| < \frac{\epsilon}{2} + \frac{\epsilon}{2} = \epsilon$$

whenever $0 < |x - a| < \delta$.

Parts (b) and (c) are left as problems. □

A *neighborhood* of a point p is an open interval $N = (a, b)$ containing p. The following result will be important in the next chapter.

Lemma 1 *If*

$$\lim_{x \to a} f(x) > 0,$$

then there is a neighborhood N of a such that

$$f(x) > 0$$

if $x \in N, x \neq a$.

Proof Let $\epsilon = |L|/2$ and choose δ such that

$$|f(x) - L| < \frac{|L|}{2}$$

if $0 < |x - a| < \delta$. Then

$$|f(x)| \geq |L| - |f(x) - L| > |L| - \frac{|L|}{2} = \frac{|L|}{2} > 0.$$ □

3.1.1 Problems

1. Prove Theorem 2(b).

2. Prove Theorem 2(c).

3. Prove that

    (a) if c is a constant, then
    $$\lim_{x \to a} c = c,$$

    (b)
    $$\lim_{x \to a} x = a, \text{ and}$$

    (c)
    $$\lim_{x \to a} x^n = a^n$$

    for every positive integer n. (*Hint*: For (c), use induction.)

4. Prove directly from the ε, δ-definition that
$$\lim_{x \to 2} x^2 = 4.$$

5. Prove directly from the ε, δ-definition that
$$\lim_{x \to 1} \frac{x}{x^2 + 1} = \frac{1}{2}.$$

6. Prove directly from the ε, δ-definition that
$$\lim_{x \to 0} \frac{x}{1 - x^2} = 0.$$

7. Prove directly from the ε, δ-definition that
$$\lim_{x \to 1} \frac{1}{x^2} = 1.$$

3.2 Limits and Sequences

The notion of limit with respect to a continuous variable can be reduced to limits of sequences. Indeed, the criterion of the following theorem could be used as the definition of limit, reducing everything to sequences.

Theorem 3 (Sequential Criterion for Limits) *Let a be a point of an interval I, and $f(x)$ a function defined on I, except possibly at the point a itself. Then*

$$\lim_{x \to a} f(x) = L \tag{1}$$

iff

$$\lim f(x_n) = L$$

for every sequence $x_n \in I, x_n \neq a$, such that $x_n \to a$.

Proof Suppose (1) holds. Let $x_n \to a$ ($x_n \neq a$). We must show that

$$f(x_n) \to L.$$

Let $\epsilon > 0$, and let δ be the number corresponding to ϵ. Since $x_n \to a$, there is an N such that

$$0 < |x_n - a| < \delta$$

whenever $n \geq N$. ($|x_n - a|$ is positive because $x_n \neq a$.) By hypothesis,

$$|f(x_n) - L| < \epsilon.$$

Therefore

$$f(x_n) \to L.$$

Conversely, assume that

$$f(x_n) \to L$$

for every sequence $x_n \in I$ such that $x_n \neq a$ and $x_n \to a$. Let $\epsilon > 0$ be given, and suppose that, for this ϵ, there is *no such number* $\delta > 0$ as described in Definition 1. In particular, for every n, the number $\delta = 1/n$ fails, so there must be a point x_n with

$$0 < |x_n - a| < \frac{1}{n}$$

for which

$$|f(x_n) - L| \geq \epsilon.$$

But if this is true, then

$$x_n \to a$$

while $f(x_n)$ cannot converge to L. This contradicts our assumption, so we conclude that $\lim_{x \to a} f(x) = L$. □

Theorems 1 and 2 follow easily from this criterion. As an example, we shall prove (b) of Theorem 2.

Let x_n be any sequence such that $x_n \to a$, $x_n \neq a$. By hypothesis,

$$f(x_n) \to L \quad \text{and} \quad g(x_n) \to M.$$

By the Limit Theorem for sequences, it follows that

$$f(x_n)g(x_n) \to LM,$$

which proves (b).

3.2.1 Problems

1. Using the sequential criterion for limits, prove the following:

    (a) Theorem 2(a),
    (b) Theorem 2(c), and
    (c) Theorem 1.

2. Prove that if

$$g(x) \leq f(x),$$

then

$$\lim_{x \to a} g(x) \leq \lim_{x \to a} f(x).$$

3. (*Sandwich Theorem*) Prove that if

$$\lim_{x \to a} g(x) = \lim_{x \to a} f(x) = L$$

and

$$g(x) \leq h(x) \leq f(x),$$

then

$$\lim_{x \to a} h(x) = L.$$

4. Prove that

$$\lim_{x \to 0} x \sin\left(\frac{1}{x}\right) = 0.$$

3.3 Continuity

The notion of a continuous function is one of the most important in Analysis.

Definition 2 Let $f(x)$ be a function defined on an interval I, and a a point of I. We say that $f(x)$ *is continuous at* a iff

$$\lim_{x \to a} f(x) = f(a).$$

A point a at which a function $f(x)$ is not continuous is called a *discontinuity* of $f(x)$.

If

$$\lim_{x \to a} f(x) = L$$

exists, but $f(a)$ is either undefined or is not equal to L, then $f(x)$ is said to have a *removable discontinuity* at a. In this case, if the function is redefined at a so that $f(a) = L$, then f will be continuous at a.

Example Since

$$\lim_{x \to 0} \frac{\sin x}{x} = 1$$

the function $f(x) = \sin x / x$, which is undefined at $x = 0$, can be defined at the origin by $f(0) = 1$ to give a function that is continuous at all points. ∎

From the Basic Limit Theorem, we have

Theorem 4 *If $f(x)$ and $g(x)$ are continuous at a, then*

(a) *$f(x) + g(x)$ and $f(x)g(x)$ are continuous at a.*

(b) *If, in addition, $g(a) \neq 0$, then $\frac{f(x)}{g(x)}$ is also continuous at a.*

From the Sequential Criterion for Limits, there follows:

Theorem 5 (Sequential Criterion for Continuity) *Let $f(x)$ be defined on I. If $a \in I$, then $f(x)$ is continuous at a iff*

$$\lim f(x_n) = f(a)$$

for every sequence $x_n \in I$ with $x_n \to a$.

Note that we may have $x_n = a$ for some, or even all, n.

Proof This follows directly from Theorem 3. □

Theorem 6 *Let $g(x)$ be continuous at a, and $f(x)$ continuous at the point $g(a)$. Then the composition*
$$(f \circ g)(x) = f(g(x))$$
of f and g is continuous at a.

Proof Let $x_n \to a$. Then, by continuity of g, $g(x_n) \to g(a)$ and so, by continuity of f, $f(g(x_n)) \to f(g(a))$. □

Definition 3 *We say that $f(x)$ is continuous on an interval I iff $f(x)$ is continuous at a for every a in I.*

Corollary 1 *The sum, product, and composition of continuous functions are continuous where they are defined. The quotient of continuous functions is continuous at all points where the denominator does not vanish.*

3.3.1 Problems

1. Which of the following functions have removable discontinuities? After all removable discontinuities are removed, which functions are continuous? At what points are they not continuous? Justify your answers.

    (a) $x^4 - x^3 + 3x + 7$
    (b) $(x^2 - 2x + 1)/(x^2 - 1)$
    (c) $\sqrt{x^2 + 1}$
    (d) $\sin(1/x)$
    (e) $x\sin(1/x)$

2. Prove that

    (a) $\lim_{x \to a} x^n = a^n$ if n is a positive integer,
    (b) a polynomial is continuous at every point, and
    (c) a rational function is continuous at every point at which its denominator does not vanish.

3. Prove Theorem 6 directly from Definition 1.

4. Prove that if $f(x)$ and $g(x)$ are continuous at c and $f(c) < g(c)$, then there is a neighborhood N of c such that
$$f(x) < g(x)$$
for all x in N.

5. Define
$$\chi_{\mathbb{Q}}(x) = \begin{cases} 0 & \text{if } x \text{ is irrational} \\ 1 & \text{if } x \text{ is rational} \end{cases}$$

   (a) At what points is $\chi_{\mathbb{Q}}(x)$ continuous?
   (b) At what points is the function
$$g(x) = \sin(x)\,\chi_{\mathbb{Q}}(x)$$
   continuous?

6. Prove that if $f(x)$ is continuous on $[a,b]$ and $f(x) = 0$ for all *rational* numbers x, then $f(x) = 0$ for all x in $[a,b]$.

3.4 Infinite Limits

The definition of infinite limits with respect to a continuous variable is analogous to the definition for sequences.

Definition 4 Let a be a point of an interval I, and $f(x)$ a function defined on I, except possibly at the point a itself. We say that
$$\lim_{x \to a} f(x) = \infty$$
iff for every positive number M there exists a positive number δ such that
$$f(x) > M$$
whenever $0 < |x - a| < \delta$ and $x \in I$.

Example We will prove that
$$\lim_{x \to \infty} (x^2 + 1) = \infty.$$

Proof Let $M > 0$. Choose $N = \sqrt{M}$. Then for $x > N$, we have
$$x^2 + 1 > x^2 > \left(\sqrt{N}\right)^2 = N.$$
∎

Theorem 7 (Sequential Criterion) We have

$$\lim_{x \to a} f(x) = \infty$$

iff

$$\lim f(x_n) = \infty$$

for every sequence $x_n \in I$ such that $x_n \to a$.

Proof See Problem 5. □

Remark Remember that if the limit is infinite, we do not say that the limit exists.

3.4.1 Problems

1. Prove directly from the definition that

$$\lim_{x \to 0} \frac{1}{x^2} = \infty.$$

2. For the following, give a definition and a sequential criterion, and prove that they are equivalent:

   (a) $\lim_{x \to \infty} f(x) = L$, and
   (b) $\lim_{x \to \infty} f(x) = \infty$.

3. Prove directly from the definition of Problem 2(b) that

$$\lim_{x \to \infty} (x^2 - x) = \infty.$$

4. Prove directly from the definition of Problem 2(a) that

$$\lim_{x \to \infty} \frac{x^2 - 1}{x^2 + 1} = 1.$$

5. Prove Theorem 7.

6. Let $f(x)$ be defined on $\mathbb{R}$, and

$$\lim_{x \to 0} f(x) = L.$$

Prove that

$$\lim_{x \to \infty} f(\frac{1}{x}) = L.$$

7. Let $f(x)$ be defined on $\mathbb{R}$, and
$$\lim_{x \to \infty} f(\frac{1}{x}) = L.$$
Is it true that
$$\lim_{x \to 0} f(x) = L?$$
If so, prove it. If not, give a counterexample and then state and prove a correct theorem.

8. Let $f(x)$ and $g(x)$ be bounded functions on $x > a$. If
$$\lim_{x \to \infty} g(x) = 0,$$
prove that
$$\lim_{x \to \infty} f(x)g(x) = 0.$$

9. Do the limits
$$\lim_{x \to \pm\infty} (|x - a| - |x - b|)$$
exist? If so, what are they?

3.5 *One-Sided Limits and Monotone Functions

One-Sided Limits

The left- and right-hand limits of functions can be defined.

Definition 5 We say that
$$\lim_{x \to a+} f(x) = L$$
iff for every $\epsilon > 0$, there exists a $\delta > 0$ such that
$$|f(x) - L| < \epsilon$$
whenever $a < x < a + \delta$.

We sometimes write
$$f(x+) = \lim_{x \to a+} f(x).$$

We refer to $f(a+)$ as the *limit of $f(x)$ from the right*, or the *right-hand limit*. In a similar manner, one defines the *left-hand limit*

$$f(a-) = \lim_{x \to a-} f(x).$$

There is a sequential criterion for right-hand limits.

Theorem 8 (Sequential Criterion) *We have*

$$\lim_{x \to a+} f(x) = L$$

iff $f(x_n) \to L$ for every sequence $x_n \to a$ such that $x_n > a$.

Proof See Problem 3. □

Theorem 9 *We have*

$$\lim_{x \to a} f(x) = L$$

iff $f(a+)$ and $f(a-)$ both exist and

$$f(a+) = f(a-) = L.$$

Proof See Problem 4. □

If $f(a+)$ and $f(a-)$ both exist, but $f(a+) \neq f(a-)$, then we say that $f(x)$ has a *jump discontinuity* at a.

Monotone Functions

We can define monotone functions as well as monotone sequences.
A function $f(x)$ defined on a set S is *increasing* iff

$$f(x) \leq f(y)$$

for all $x, y \in S$ with $x \leq y$. It is *decreasing* iff

$$f(x) \geq f(y)$$

for all $x, y \in S$ with $x \leq y$.

A function that is either increasing or decreasing is called a *monotone function*.

As with sequences, we call $f(x)$ *strictly increasing* iff $f(x) < f(y)$ when $x < y$, and *strictly decreasing* iff $f(x) > f(y)$ whenever $x < y$.

3.5 *One-Sided Limits and Monotone Functions

Again, as with sequences, some writers use the term *nondecreasing* for what we have called *increasing*, and *nonincreasing* for what we have called *decreasing*, reserving the term *increasing* to mean what we have called *strictly increasing*.

Examples The functions $f(x) = x$ and $g(x) = e^x$ are strictly increasing on the whole line $\mathbb{R}$.

The function $f(x) = x^2$ is strictly increasing on the interval $[0, \infty)$ and strictly decreasing on $(-\infty, 0]$.

The function $f(x) = \cos x$ is strictly *increasing* on the interval $I_n = [n\pi, \pi(n+1)]$ if n is *odd* and strictly *decreasing* on I_n if n is *even*.

The function

$$H(x) = \begin{cases} 1 & \text{for } x \geq 0 \\ 0 & \text{for } x < 0 \end{cases}$$

is increasing on the whole line $\mathbb{R}$, but not strictly increasing. ■

The important fact about monotone functions is the following:

Theorem 10 *If $f(x)$ is monotone, then $f(x+)$ and $f(x-)$ exist at all points. Thus, monotone functions can have only jump discontinuities.*

Proof Assume that $f(x)$ is increasing. We shall prove that $f(a-)$ exists. Define

$$L = \sup \{f(x) : x < a\}.$$

We claim that

$$f(a-) = L.$$

If L is finite, let $\epsilon > 0$. Choose $c < A$ such that

$$L - \epsilon < f(c) \leq L$$

and let $\delta = a - c$. If $a - \delta = c < x < a$, then

$$L - \epsilon < f(c) \leq f(x) \leq L.$$

Existence of $f(a+)$ follows by considering the function $f(-x)$.

The case $L = \infty$ is deferred to Problem 5. □

3.5.1 Problems

1. Give a precise definition of the statement
$$\lim_{x \to a-} f(x) = L$$
in terms of ϵ and δ, and give the corresponding sequential criterion.

2. Give a precise definition of the statement
$$\lim_{x \to a-} f(x) = \infty$$
in terms of ϵ and δ, and give the corresponding sequential criterion.

3. Prove Theorem 8.

4. Prove Theorem 9.

5. Prove Theorem 10 when $L = \infty$.

3.6 The Intermediate Value Theorem

We now consider the first of two major theorems on continuous functions. Both are consequences of the property of Continuity. Each one asserts the *existence of a real number* where something happens. The Principle of Continuity is needed to ensure that these points are actually there.

Intuitively, a continuous function is one whose graph is an unbroken curve. Such a curve cannot cross from a point below the axis to a point above the axis, without passing through the axis at some point. The precise statement of this notion is called the *Intermediate Value Theorem*. It is due to Bolzano.

Theorem 11 (Intermediate Value Theorem) *Let $f(x)$ be continuous on the closed interval $[a, b]$, and suppose that*
$$f(a) < 0 < f(b).$$
Then there exists a point c, $a < c < b$, such that
$$f(c) = 0.$$

Proof The proof is by the *Method of Bisection*. Let m be the midpoint of $[a, b]$. It is possible that $f(m) = 0$. If so, then m is the desired point. If not (as is more likely), then $f(x)$ has opposite signs at the endpoints of one of the two intervals $[a, m]$ and $[m, b]$. Denote this interval by
$$I_1 = [a_1, b_1].$$

(Thus, I_1 is $[a, m]$ if $f(m) > 0$, while $I_1 = [m, b]$ if $f(m) < 0$.)

Continue this process by bisecting I_1, and choosing $I_2 = [a_2, b_2]$ to be whichever half of I_1 satisfies

$$f(a_2) < 0 < f(b_2).$$

This process may be continued indefinitely to give a nested sequence of intervals $I_n = [a_n, b_n]$ satisfying

$$f(a_n) < 0 < f(b_n).$$

(The only thing that could prevent this would be if $f(m_n) = 0$ at the midpoint m_n of some I_n. In this case, however, that midpoint may be taken to be the desired point.) The length of I_n tends to zero, so by Cantor's Principle, there is a unique point c in the intersection of the intervals I_n; that is,

$$a_n < c < b_n$$

for all n.

Now $b_n \to c$ and $0 < f(b_n)$, so by continuity,

$$f(c) = \lim f(b_n) \geq 0.$$

Similarly, $f(b_n) < 0$, and

$$f(c) = \lim f(a_n) \leq 0.$$

So

$$0 \leq f(c) \leq 0,$$

which implies that $f(c) = 0$. □

Corollary 2 *If $f(x)$ is continuous on an interval $[a, b]$, then $f(x)$ takes on every value between $f(a)$ and $f(b)$ at some point in $[a, b]$.*

Proof See Problem 1. □

A function such that whenever $f(a) < C < f(b)$, there exists a point c between a and b such that $f(c) = C$ is said to have the *Intermediate Value Property*. Such a function need not necessarily be continuous. For example, the function

$$f(x) = \sin\left(\frac{1}{x}\right)$$

with $f(0) = 0$ is not continuous, but has this property.

3.6.1 Problems

1. Prove Corollary 2.

2. Prove that there is a real solution of the equation
$$x = e^{-x}.$$
(*Hint*: Draw a picture.)

3. Let $f(x)$ be continuous on $\mathbb{R}$, and suppose that $f(x)$ is a rational number for every x. Prove that $f(x)$ is constant.

4. Let
$$f(x) = \begin{cases} \sin(1/x) & \text{if } x \neq 0 \\ 0 & \text{if } x = 0. \end{cases}$$
Prove that $f(x)$ has the Intermediate Value Property, although $f(x)$ is not continuous.

5. Prove that if $f(x)$ and $g(x)$ are continuous on $[a,b]$, $f(a) < g(a)$, and $f(b) > g(b)$, then there exists a point c, $a < c < b$, such that $f(c) = g(b)$.

6. Prove that a polynomial of odd degree has a real root.

7. Prove that a positive real number a has a positive real nth root. (*Hint*: Consider $p(x) = x^n - a$.)

8. If f is one to one and continuous on $[a,b]$, then f is strictly monotone on $[a,b]$.

9. Show by an example that the Intermediate Value Theorem is false if we consider only rational numbers.

3.7 The Extreme Value Theorem

The second major theorem on continuous functions is the *Extreme Value Theorem*.

A function $f(x)$ defined on a set D is said to be *bounded above* on D iff there is an M such that
$$f(x) \leq M$$
for all $x \in D$. If there is an M such that
$$M \leq f(x),$$
then $f(x)$ is said to be *bounded below* on D.

3.7 The Extreme Value Theorem

We say that $f(x)$ is *bounded* on D if $f(x)$ is bounded above and below; that is, if there is an M such that

$$|f(x)| \leq M$$

for all $x \in D$.

Theorem 12 *A continuous function on a closed, bounded interval is bounded.*

Proof Let $f(x)$ be continuous on $[a,b]$. Suppose that $f(x)$ is not bounded above on $[a,b]$. Then for every positive integer n, there is a point x_n in $[a,b]$ such that

$$f(x_n) > n$$

since otherwise n would be an upper bound of $f(x)$. The sequence x_n is a bounded sequence, since it is contained in the bounded interval $[a,b]$. By the Bolzano–Weierstrass Theorem, there is a subsequence x_{n_k} of x_n that converges to a point c. Since

$$a \leq x_{n_k} \leq b$$

and $x_{n_k} \to c$, we must have $a \leq c \leq b$, so that c is in $[a,b]$. Since f is continuous at c, $f(x_{n_k}) \to f(c)$. But by construction, $f(x_{n_k}) \to \infty$, which is a contradiction. Thus, $f(x)$ is bounded above.

Now, by what was just proved, the function $g(x) = -f(x)$ is bounded above, and hence $f(x)$ is bounded below. □

Let $f(x)$ be defined on a set D. We say that $f(x)$ *attains a maximum at a point c of D* iff

$$f(x) \leq f(c)$$

for all points x of D.

Theorem 13 (Extreme Value Theorem) *A continuous function on a closed, bounded interval $[a,b]$ attains a maximum at some point of $[a,b]$.*

Proof Let $f(x)$ be continuous on $[a,b]$. The number

$$M = \sup\{f(x) : a \leq x \leq b\}$$

is finite by Theorem 12. We need to show that $f(x)$ actually *attains* the value M at some point of $[a,b]$.

Since the supremum of a set is the limit of a sequence of points in the set, there exists a sequence of points x_n in $[a,b]$ such that

$$f(x_n) \to M.$$

By the Bolzano–Weierstrass Theorem, x_n has a subsequence x_{n_k} converging to a point c in $[a,b]$. Since x_{n_k} is a subsequence of x_n,

$$f(x_{n_k}) \to M.$$

But by continuity of f,

$$f(x_{n_k}) \to f(c).$$

Since limits are unique, we must have $f(c) = M$. $\square$

We say that $f(x)$ *attains a minimum at a point c of D* iff

$$f(x) \geq f(c)$$

for all points x of D.

Since the function $f(x)$ will have a minimum at any point where the function $g(x) = -f(x)$ has a maximum, we also obtain:

Corollary 3 *A continuous function on a closed, bounded interval $[a,b]$ attains a minimum at a point of $[a,b]$.*

3.7.1 Problems

1. Let $f(x)$ be continuous on a closed bounded interval $I = [a,b]$. Suppose that for every x in I there exists a y in I such that

$$|f(y)| \leq \frac{1}{2}|f(x)|.$$

    Prove that there is a point c in I such that $f(c) = 0$.

    Give counterexamples to show that this is not true if I is not closed or if I is infinite.

2. Let $f(x)$ be continuous on a closed bounded interval $[a,b]$. Prove that the range of $f(x)$ is also a closed, bounded interval.

3. Let $C : (x(t), y(t))$, $a \leq t \leq b$ be a continuous curve in the plane. Prove that there is a point on C that is closest to the origin.

4. A function f has a *local maximum (minimum)* at a point c iff there is a neighborhood N of c such that the maximum (minimum) of f on N occurs at c. Prove that if f is continuous on $[a,b]$ and has no interior local maximum or minimum, then f is strictly monotone on $[a,b]$.

5. Show by examples that neither 'continuous' nor 'closed' nor 'bounded' can be omitted from the hypothesis of the Extreme Value Theorem.

6. Show by an example that the Extreme Value Theorem is false if we consider only rational numbers.

3.8 Supplementary Problems

1. Define
$$\chi_\mathbb{Q}(x) = \begin{cases} 0 & \text{if } x \text{ is irrational} \\ 1 & \text{if } x \text{ is rational.} \end{cases}$$

   Does
$$\lim_{x \to 0} x\chi_\mathbb{Q}(x)$$
   exist? If so what is it? Prove that your answer is correct.

2. (*Cauchy's Criterion*) Prove that $\lim_{x \to a} f(x)$ exists iff
$$\lim_{x,y \to a} [f(x) - f(y)] = 0;$$
   that is, for every $\epsilon > 0$, there exists a $\delta > 0$ such that
$$|f(x) - f(y)| < \epsilon$$
   whenever $|x - a| < \delta$, and $|y - a| < \delta$.

*3. Define
$$\psi(x) = \begin{cases} 0 & \text{if } x \text{ is irrational} \\ 1/q & \text{if } x = p/q \text{ in lowest terms.} \end{cases}$$

   At what points is $\psi(x)$ continuous?

*4. Let $f(x)$ be continuous on $\mathbb{R}$ and suppose that
$$f(r) = r^2$$
   for every *rational* number r. Is it true that $f(x) = x^2$ for every *real* number x?

*5. Let S be a subset of the real numbers, and x a point. Define the *distance from x to S* to be
$$d(x, S) = \inf \{|x - s| : s \in S\}.$$
   Prove that $d(x, S)$ is continuous.
   (*Hint*: Prove that $|d(x, S) - d(y, S)| \leq |x - y|$.)

6. Let $f(x)$ be defined and continuous on $-\infty < x < \infty$, and suppose that
$$f(x + y) = f(x) + f(y).$$
   Prove that $f(x) = cx$ for some constant c.

*7. (*Oscillation of a function*) Let $f(x)$ be defined and bounded on $[a,b]$. If $I = [c,d] \subset [a,b]$, define the *oscillation of f over I* to be

$$\omega(I) = \sup\{|f(x) - f(y)| : x, y \in I\}.$$

For each point c in $[a,b]$, define the *oscillation of f at c* to be

$$\omega(c) = \lim_{h \to 0+} \omega\left([c-h, c+h] \cap [a,b]\right).$$

(a) Prove that $\omega(I)$ and $\omega(c)$ exist.

(b) Prove that f is continuous at c iff $\omega(c) = 0$.

8. Do the limits

$$\lim_{x \to 0\pm} x\sqrt{1 + \frac{1}{x^2}}$$

exist and if so, what are they?

9. If f is continuous on (a,b), and if for each point of c of (a,b) there is an open interval I containing c such that f is increasing on I, prove that f is increasing on (a,b).

(*Hint*: Let $c = \sup\{x : f \text{ is increasing on } [a,x]\}$.) Prove that $c = b$.

10. (*Brouwer Fixed-Point Theorem*) A point c is called a *fixed point* of a function $f(x)$ iff $f(c) = c$.

Let $I = [0,1]$, and let $f(x)$ be a continuous function mapping I into I; that is, $0 \le f(x) \le 1$ for all x with $0 \le x \le 1$. Prove that $f(x)$ has a fixed point in I.

(*Hint*: Draw a picture. Let $g(x) = x - f(x)$. Use the Intermediate Value Theorem.)

*11. (*Alternate proof of the Intermediate Value Theorem*) Let $f(x)$ be continuous on $[a,b]$ and assume that

$$f(a) < 0 < f(b).$$

Define

$$c = \sup\{x : a \le x \le b \text{ and } f(t) < 0 \text{ for } a \le t < x\}$$

and prove that $f(c) = 0$.

12. Prove that if $f(x)$ is *monotone* and has the Intermediate Value Property, then $f(x)$ is continuous.

Chapter 4
Derivatives

4.1 The Derivative

Let $f(x)$ be a function defined on the open interval $a < x < b$, and let c be a point of (a, b).

Definition 1 The *derivative of $f(x)$ at c* is the number
$$f'(c) = \lim_{h \to 0} \frac{f(c+h) - f(c)}{h}.$$
If this limit exists for a point c, we say that f is *differentiable* at c.

Geometrically, the derivative is the slope of the tangent line to the graph of $f(x)$ at $x = a$.

If f is differentiable at every point of an interval (a, b), then $f'(x)$ defines a new function of x whose value at any point x is the number $f'(x)$. This new function is called the *derivative* of $f(x)$.

Remark The definition may also be written as
$$f'(c) = \lim_{x \to c} \frac{f(x) - f(c)}{x - c}.$$

Theorem 1 *If $f(x)$ is differentiable at c, then $f(x)$ is continuous at c.*

Proof We have
$$f(x) = f(c) + \left[\frac{f(x) - f(c)}{x - c}\right](x - c).$$

Hence,
$$\lim_{x \to c} f(x) = \lim_{x \to c} f(c) + \lim_{x \to c} \left[\frac{f(x) - f(c)}{x - c} \right] \lim_{x \to c} (x - c)$$
$$= f(c) + f'(c) \cdot 0 = f(c). \qquad \square$$

Hence, *a differentiable function is always continuous.* The converse is not true, as the following examples show.

Example 1 The derivative of the function $f(x) = |x|$ does not exist at $x = 0$. For
$$\frac{|0 + h| - |0|}{h} = \frac{|h|}{h} = \begin{cases} 1 & \text{if } h > 0 \\ -1 & \text{if } h < 0. \end{cases}$$
The limit of this expression does not exist as $h \to 0$.

Geometrically, the derivative of $|x|$ fails to exist because the graph has a sharp corner on it, so that *no tangent line exists*. ∎

Example 2 The derivative of the function $f(x) = x^{2/3}$ does not exist at $x = 0$. For
$$\lim_{h \to 0} \frac{(0+h)^2 - 0^{2/3}}{h} = \lim_{h \to 0} \frac{h^{2/3}}{h} = \lim_{h \to 0} \frac{1}{h^{2/3}} = \infty.$$
Thus again, the limit does not exist as $h \to 0$, this time because the expression becomes infinite.

Geometrically, the derivative fails to exist because the graph has a vertical tangent at $x = 0$, which has infinite slope. ∎

Remark In these examples, the function is everywhere continuous, but the derivative fails to exist at a single point. There are, however, examples of functions that are *continuous at every point, but whose derivative exists at no points whatsoever*. The first such example was given by Weierstrass. These functions are a bit too complicated to be described at this time.

It is also *not* true that the derivative of a function is always continuous, as the next example shows.

Example 3 The function
$$f(x) = \begin{cases} x^2 \sin(1/x) & \text{if } x \neq 0 \\ 0 & \text{if } x = 0 \end{cases}$$

is *differentiable at every point x but its derivative is not continuous at $x = 0$.*
For if $x \neq 0$, then

$$f'(x) = 2x \sin\left(\frac{1}{x}\right) - \cos\left(\frac{1}{x}\right).$$

Thus

$$\lim_{h \to 0} f'(x)$$

fails to exist because the second term oscillates wildly.

This formula does not hold for $x = 0$. (Why not?) Instead, we must compute $f'(0)$ from the definition. We have

$$f'(0) = \lim_{h \to 0} \frac{f(0+h) - f(0)}{h}$$
$$= \lim_{h \to 0} \left(\frac{1}{h}\right) h^2 \sin\left(\frac{1}{h}\right) = \lim_{h \to 0} h \sin\left(\frac{1}{h}\right) = 0.$$

Thus $f'(0)$ also exists, so that $f(x)$ is differentiable at every point x. ∎

4.1.1 Problems

1. Is the function defined by

$$f(x) = x \sin\left(\frac{1}{x}\right)$$

with $f(0) = 0$, continuous at $x = 0$? Is it differentiable at $x = 0$?

2. Show that the function

$$f(x) = \begin{cases} x^2 \sin(1/x^2) & \text{if } x \neq 0 \\ 0 & \text{if } x = 0 \end{cases}$$

is differentiable at all points, but its derivative $f'(x)$ is not bounded near $x = 0$.

3. Let $L(x)$ be a function defined on $x > 0$, such that

$$L(xy) = L(x) + L(y)$$

for all $x, y > 0$.

(a) Prove that $L(1) = 0$.

(b) Prove that $L(1/x) = -L(x)$.

(c) Prove that if $L'(1)$ exists, then $L'(x)$ exists for all $x > 0$, and
$$L'(x) = L'(1)\frac{1}{x}.$$

Hint: Show that
$$\frac{L(x+h) - L(x)}{h} = \frac{1}{x}\left[\frac{L(1+\frac{h}{x})}{(h/x)}\right].$$

4.2 Rules for Derivatives

The rules for computing derivatives are contained in the following theorems.

Theorem 2

(a) *If c is a constant, then*
$$Dc = 0.$$

(b) *If n is a natural number, then*
$$Dx^n = nx^{n-1}.$$

Proof See Problem 1. □

Theorem 3 (Linearity) *If $f(x)$ and $g(x)$ are differentiable at a, and c is a constant, then $f(x) + g(x)$, and $cf(x)$ are also differentiable at a and have the derivatives*
$$f'(a) + g'(a)$$
and
$$cf'(a),$$
respectively.

Proof See Problem 2. □

Theorem 4 (Product Rule) *If $f(x)$ and $g(x)$ are differentiable at a, then $G(x) = f(x)g(x)$ is also differentiable at a and has derivative*
$$G'(a) = f'(a)g(a) + f(a)g'(a).$$

Proof We have
$$\lim_{h \to 0} \frac{f(a+h)g(a+h) - f(a)g(a)}{h}$$
$$= \lim_{h \to 0} \left\{ f(a+h) \frac{g(a+h) - g(a)}{h} + \frac{f(a+h) - f(a)}{h} g(x) \right\}$$
$$= \lim_{h \to 0} f(a+h) \lim_{h \to 0} \left[\frac{g(a+h) - g(a)}{h} \right] + \lim_{h \to 0} \left[\frac{f(a+h) - f(a)}{h} \right] g(a)$$
$$= f'(a)g(a) + f(a)g'(a)$$

where we have used continuity of f at a. □

Theorem 5 (Quotient Rule) *If $f(x)$ and $g(x)$ are differentiable at a, and $g(a) \neq 0$, then*
$$F(x) = \frac{f(x)}{g(x)}$$
is also differentiable at a and has derivative
$$F'(a) = \frac{f'(a)g(a) - f(a)g'(a)}{g(a)^2}.$$

Proof See Problem 3. □

Theorem 6 (Chain Rule) *If $f(x)$ is differentiable at a and $g(x)$ is differentiable at $b = f(a)$, then $F(x) = g(f(x))$ is differentiable at a, and*
$$F'(a) = g'(f(a))\, f'(a).$$

The proof will require the following Lemma.

Lemma 1 *Let $f(x)$ be differentiable at a. Then*
$$f(a+h) = f(a) + f'(a)h + h\, \varepsilon(h) \tag{1}$$
where
$$\lim_{h \to 0} \varepsilon(h) = 0.$$

Proof of Lemma 1 Define
$$\varepsilon(h) = \frac{f(a+h) - f(a)}{h} - f'(a).$$

Then (1) holds and
$$\lim_{h \to 0} \varepsilon(h) = f'(a) - f'(a) = 0. \qquad \square$$

Proof of Theorem 6 Let $b = f(a)$. By Lemma 1, we have
$$F(a+h) - F(a) = g(b+k) - g(b) = g'(b)k + k\,\varepsilon_2(k)$$
where
$$k = f(a+h) - f(a) = f'(a)h + h\,\varepsilon_1(h).$$
Since k tends to zero as h tends to zero, we also have
$$\lim_{h \to 0} \varepsilon_2(k) = 0.$$
Thus
$$\lim_{h \to 0} \frac{F(a+h) - F(a)}{h} = \lim_{h \to 0} [g'(b) + \varepsilon_2(k)] \frac{k}{h}$$
$$= \lim_{h \to 0} [g'(b) + \varepsilon_2(k)] \lim_{h \to 0} \left[\frac{f(a+h) - f(a)}{h} \right]$$
$$= f'(a)g'(b) = g'(f(a))\,f'(a). \qquad \square$$

4.2.1 Problems

1. Prove that

   (a) $Dc = 0$ if c is constant.
   (b) $Dx = 1$.
   (c) $Dx^n = nx^{n-1}$ for every natural number n.

   This proves Theorem 2.

2. Prove Theorem 3.

3. Prove Theorem 5.

4. Prove the following converse of Lemma 1. *If there exists a number A, such that*
$$f(a+h) = f(a) + Ah + h\varepsilon(h)$$
where
$$\lim_{h \to 0} \varepsilon(h) = 0,$$
then $f(x)$ is differentiable at a, and $f'(a) = A$.

4.3 The Critical Point Theorem

The Critical Point Theorem is the familiar tool for finding maxima and minima. We begin our discussion with a Lemma.

Lemma 2 *If $f'(c)$ exists and is positive, then there is a neighborhood of c on which*

$$f(x) > f(c) \quad \text{if} \quad x > c$$
$$f(x) < f(c) \quad \text{if} \quad x < c.$$

Proof If

$$\lim_{x \to c} F(x) > 0,$$

then the function $F(x)$ must be positive in a neighborhood of c. If $F(x)$ is the function

$$F(x) = \frac{f(x) - f(c)}{x - c},$$

then by Lemma 1 of Section 3.1,

$$\lim_{x \to c} F(x) = f'(c) > 0.$$

Hence, on a neighborhood N of c

$$\frac{f(x) - f(c)}{x - c} > 0. \tag{2}$$

The numerator and denominator of (2) must therefore have the same sign on N. Thus

$$f(x) - f(c) > 0 \quad \text{if} \quad x - c > 0$$
$$f(x) - f(c) < 0 \quad \text{if} \quad x - c < 0.$$

This is just the statement of Lemma 2. □

Theorem 7 (Critical Point Theorem) *If the function $f(x)$ has a maximum at an interior point c of $[a, b]$ and if $f'(c)$ exists, then $f'(c) = 0$.*

Proof By Lemma 2, if $f'(c) > 0$, then $f(x) > f(c)$ for $x > c$, so c is not a maximum. Similarly, if $f'(c) < 0$, then $f(x) > f(c)$ for $x < c$, so c is again not a maximum. Therefore, since $f'(c)$ can be neither positive nor negative, it must be zero, if it exists. □

The following is a simple, but interesting, consequence of the Critical Point Theorem.

Theorem 8 (Intermediate Value Theorem for Derivatives) *Let $f(x)$ be differentiable on (a, b). Then its derivative $f'(x)$ has the intermediate value property; that is, if $a < \alpha < \beta < b$, and*

$$f'(\alpha) > 0 > f'(\beta),$$

then there is a point c between α and β where

$$f'(c) = 0.$$

Proof The function f is continuous since it is differentiable. It therefore has a maximum on the interval $[\alpha, \beta]$. By Lemma 2, $f(x) > f(\alpha)$ to the right of α and $f(x) > f(\beta)$ to the left of β. The maximum therefore cannot occur at either α or β. Hence, by the Critical Point Theorem, it occurs at a point c between α and β where we must then have

$$f'(c) = 0.$$

□

Remark This theorem is interesting because the function $f'(x)$ does not have to be continuous on (a, b), as we saw in Example 3 of Section 4.1.

4.3.1 Problems

1. Prove that the function

   $$f(x) = \frac{x}{2} + x^2 \sin\left(\frac{1}{x}\right)$$

   has $f'(0) > 0$ but is not increasing in any neighborhood of $x = 0$.
   Why doesn't this contradict Lemma 2?

2. Find the maximum and minimum points of $f(x) = x^{2/3}$ on the interval $-1 \leq x \leq 1$. Do these occur at critical points? Discuss the relationship of this example to Theorem 7.

4.4 The Mean Value Theorem

The important result known as the Mean Value Theorem can be deduced from one of its special cases.

Theorem 9 (Rolle's Theorem) *Let $f(x)$ be continuous on the closed, finite interval $[a,b]$, and differentiable on the open interval (a,b), and suppose that*
$$f(a) = f(b) = 0.$$
Then there is a point c on (a,b) where
$$f'(c) = 0.$$

Proof By the Extreme Value Theorem, $f(x)$ has a maximum and a minimum on $[a,b]$. If *both* of these occur at an endpoint of $[a,b]$, then the function $f(x)$ must be identically zero, since its maximum and minimum values are both zero. In this case, $f'(c) = 0$ at *all* points of (a,b).

Otherwise, $f(x)$ has an extremum at an interior point c. By the Critical Point Theorem, $f'(c) = 0$. □

Theorem 10 (Mean Value Theorem) *Let $f(x)$ be continuous on the closed, finite interval $[a,b]$, and differentiable on the open interval (a,b). Then there is a point c on (a,b) where*
$$f'(c) = \frac{f(b) - f(a)}{b - a}. \tag{3}$$

Proof The Mean Value Theorem follows by applying Rolle's Theorem to the function obtained by subtracting from $f(x)$ the equation of the straight line passing through the points $(a, f(a))$ and $(b, f(b))$.

To be precise, let
$$m = \frac{f(b) - f(a)}{b - a}$$
and
$$\ell(x) = m(x - a) + f(a).$$
Then $\ell(a) = f(a)$ and $\ell(b) = f(b)$. The function
$$g(x) = f(x) - \ell(x)$$
therefore satisfies the hypotheses of Rolle's Theorem. Hence, there is a point c where
$$g'(c) = f'(c) - \ell'(c) = f'(c) - m = 0.$$
Hence,
$$f'(c) = m,$$
which is (3). □

The Mean Value Theorem has several important consequences.

Corollary 1 *If*
$$f'(x) = 0$$
for all x in (a,b), then there is a constant C such that
$$f(x) = C$$
for all x in (a,b).

Proof Let $a < x < y < b$. By the Mean Value Theorem, there is a point c between x and y for which
$$\frac{f(y) - f(x)}{y - x} = f'(c).$$
By hypothesis, $f'(c) = 0$, regardless of where c is. Hence,
$$f(y) - f(x) = 0$$
or
$$f(y) = f(x).$$
But x and y were any two points in (a,b), so this says that $f(x)$ has the same value at all points of (a,b); that is, $f(x)$ is a constant. □

Corollary 2 *If*
$$f'(x) > 0$$
for all x in (a,b), then $f(x)$ is strictly increasing on (a,b).

Proof Let $a < x < y < b$. By the Mean Value Theorem, there is a point c between x and y for which
$$\frac{f(y) - f(x)}{y - x} = f'(c).$$
By hypothesis, $f'(c) > 0$, regardless of where c is. Hence,
$$f(y) - f(x) > 0$$
or
$$f(y) > f(x),$$
which says that $f(x)$ is strictly increasing. □

4.4 The Mean Value Theorem

Theorem 11 (The Second Derivative Test) *Let $f(x)$ be differentiable on (a,b), and c a point of (a,b) where*

$$f'(c) = 0.$$

(a) *If $f''(c) < 0$, then c is a local maximum.*

(a) *If $f''(c) > 0$, then c is a local minimum.*

Proof We shall prove only (a). If $f''(c) < 0$, then, since $f''(x)$ is the derivative of $f'(x)$, there is a neighborhood N of c on which

$$f'(x) < 0 \quad \text{for} \quad x > c$$

and

$$f'(x) > 0 \quad \text{for} \quad x < c.$$

Thus on N, $f(x)$ is increasing for $x < c$, and decreasing for $x > c$. It follows that c is a maximum of f on N. □

The following generalization of the Mean Value Theorem is often useful. It reduces to the Mean Value Theorem if $g(x) = x$.

Theorem 12 (Cauchy's Mean Value Theorem) *Let $f(x)$ and $g(x)$ be continuous on $[a,b]$ and differentiable on (a,b). Then there exists a point c, $a < c < b$, with*

$$[f(b) - f(a)]\, g'(c) = [g(b) - g(a)]\, f'(c).$$

Proof See Problem 4. □

4.4.1 Problems

1. Assume that $f(x)$ is defined for all points in a neighborhood of the point a, and that $f'(x)$ exists, except possibly for $x = a$.

   Prove that if

   $$\lim_{x \to a} f'(x) = L$$

   exists, then $f'(a)$ exists, and

   $$f'(a) = L.$$

   (*Hint:* Use the Mean Value Theorem.)

2. Define, for α and β positive,

$$f(x) = \begin{cases} x^\alpha \sin(1/x^\beta) & \text{if } x \neq 0 \\ 0 & \text{if } x = 0. \end{cases}$$

For what values of α and β is

   (a) $f(x)$ differentiable at all points?
   (b) $f'(x)$ bounded on $[-1, 1]$?
   (c) $f'(x)$ continuous at all points?

3. Prove that if $f(x)$ and $g(x)$ are differentiable functions such that $f(a) \leq g(a)$ and

$$f'(x) \leq g'(x)$$

for $a \leq x \leq b$, then

$$f(x) \leq g(x)$$

for $a \leq x \leq b$.

4. Prove Cauchy's Mean Value Theorem by considering the function

$$F(x) = [f(b) - f(a)][g(x) - g(a)] - [g(b) - g(a)][f(x) - f(a)].$$

5. Use Rolle's Theorem to show that the polynomial

$$p(x) = x^3 - 3x + b$$

cannot have more than one root in the interval $-1 \leq x \leq 1$, regardless of what b is.

6. Prove that if $p > 0$, then there exists a constant C_p such that for $x > 0$,

$$x^p \leq C_p e^x.$$

(*Hint:* Examine the derivative of $x^p e^{-x}$.)

7. Let $p > 0$. Prove that there exists a constant C_p such that

$$\log x \leq C_p x^p$$

for $x \geq 1$, and

$$|\log x| \leq \frac{C_p}{x^p}$$

for $0 < x \leq 1$.

8. Assume that $f(x)$ is differentiable on $x > 0$ and that

$$\lim_{x \to \infty} f'(x) = 0.$$

Prove that

$$\lim_{x \to \infty} f(x+1) - f(x) = 0.$$

9. Show that the second derivative test is inconclusive when the second derivative vanishes.

(*Hint:* Give examples where $f'(c) = f''(c) = 0$, but f has, respectively, a maximum, a minimum, and no extremum at c.)

4.5 *L'Hospital's Rule

L'Hospital's Rule for the Case $\frac{0}{0}$

We shall give two proofs of L'Hospital's familiar rule for indeterminate limits.

Theorem 13 *Let $f(x)$ and $g(x)$ be differentiable for $0 < x < a$ with $g'(x) > 0$. If*

$$\lim_{x \to 0+} f(x) = \lim_{x \to 0+} g(x) = 0,$$

and if

$$\lim_{x \to 0+} \frac{f'(x)}{g'(x)} = L$$

exists, then

$$\lim_{x \to 0+} \frac{f(x)}{g(x)} = L.$$

Proof 1 Let $x > 0$. Apply Cauchy's Mean Value Theorem to the interval $[0, x]$ to obtain

$$\frac{f(x) - f(0)}{g(x) - g(0)} = \frac{f(x)}{g(x)} = \frac{f'(c)}{g'(c)}$$

for some $c = c(x)$ with $0 < c < x$. Since

$$\lim_{x \to 0+} c(x) = 0$$

we get

$$\lim_{x \to 0+} \frac{f(x)}{g(x)} = \lim_{x \to 0+} \frac{f'(c(x))}{g'(c(x))} = L.$$

Note that the assumptions *imply* that $g(x) > 0$. (Why?) □

The following proof avoids Cauchy's Mean Value Theorem.

Proof 2 Let $\epsilon > 0$, and choose $\delta > 0$ so that

$$L - \epsilon < \frac{f'(x)}{g'(x)} < L + \epsilon$$

whenever $0 < x < \delta$. Hence,

$$(L - \epsilon) g'(x) < f'(x) < (L + \epsilon) g'(x).$$

Since we may as well assume that $f(0) = g(0) = 0$, Problem 3 of Section 4.4 implies that

$$(L - \epsilon) g(x) < f(x) < (L + \epsilon) g(x),$$

which is the desired result. □

Example 1 Consider the limit

$$\lim_{x \to 0} \frac{\sin x}{x}.$$

By Theorem 13,

$$\lim_{x \to 0} \frac{\sin x}{x} = \lim_{x \to 0} \frac{\cos x}{1} = 1.$$

■

L'Hospital's Rule for the Case $\frac{\infty}{\infty}$

The case of an indeterminate limit of the form ∞/∞ does not follow from the case $0/0$, but needs a separate proof.

Theorem 14 Let $f(x)$ and $g(x)$ be differentiable for $0 < x < a$ with $g'(x) < 0$. If

$$\lim_{x \to 0+} f(x) = \lim_{x \to 0+} g(x) = \infty,$$

and if

$$\lim_{x \to 0+} \frac{f'(x)}{g'(x)} = L$$

exists, then

$$\lim_{x \to 0+} \frac{f(x)}{g(x)} = L.$$

Proof Let $x > a > b$. By Cauchy's Mean Value Theorem applied to the interval $[x, a]$ there is a $c = c(x)$ with $x < c < a$ such that

$$\frac{f'(c)}{g'(c)} = \frac{f(x) - f(a)}{g(x) - g(a)} = \frac{f(x)}{g(x)} \left\{ \frac{1 - \frac{f(a)}{f(x)}}{1 - \frac{g(a)}{g(x)}} \right\}.$$

Since

$$\lim_{x \to 0+} c(x) = 0$$

we get

$$\lim_{x \to 0+} \frac{f(x)}{g(x)} = \lim_{x \to 0+} \frac{f'(c(x))}{g'(c(x))} \left\{ \frac{\lim_{x \to 0+} \left(1 - \frac{g(a)}{g(x)}\right)}{\lim_{x \to 0+} \left(1 - \frac{f(a)}{f(x)}\right)} \right\} = L.$$

(Again, the assumptions *imply* that $g(x) > 0$.) $\square$

Example 2 Consider the limit

$$\lim_{x \to 0} (x^\alpha \log x) = \lim_{x \to 0} \frac{\log x}{x^{-\alpha}}$$

where $\alpha > 0$. By Theorem 14,

$$\lim_{x \to 0} \frac{\log x}{x^{-\alpha}} = \lim_{x \to 0} \frac{1/x}{-\alpha x^{-\alpha - 1}} = -\lim_{x \to 0} \left(\frac{x^\alpha}{\alpha}\right) = 0.$$

Interpretation This says that $\log x$ *blows up as* $x \to 0$ *more slowly than any positive reciprocal power of* x.

4.5.1 Problems

1. Find *and interpret* the following limits for $p > 0$.

   (a) $\lim_{x \to \infty} \frac{x^p}{e^x}$

   (b) $\lim_{x \to \infty} \frac{\log x}{x^p}$

   (c) $\lim_{x \to 0} x^p e^{-1/x}$

   (d) $\lim_{x \to 0} x^{1/x}$

2. Show that Theorem 13 holds in the following cases:

   (a) $\lim_{x \to a+}$ for any a

   (b) $\lim_{x \to a-}$ for any a

   (c) $\lim_{x \to a}$ for any a

   (d) $\lim_{x \to \pm \infty}$

3. (*L'Hospital*) Find

$$\lim_{x \to a} \frac{\sqrt{2a^3 x - x^4} - a \sqrt[3]{a^2 x}}{a - \sqrt[4]{ax^3}}.$$

   (*Hint*: The scaling substitution $x = at$ will reduce the problem to the case $a = 1$.)

4. "By L'Hospital's Rule,

$$\lim_{x \to 0} \frac{2+x}{1+x} = \lim_{x \to 0} \frac{1}{1} = 1."$$

   Comment on this assertion.

5. Discuss L'Hospital's Rule for the case where $L = \infty$.

6. Why is $g(x) > 0$ in Proof 1?

7. Prove Theorem 14 without using Cauchy's Mean Value Theorem.

4.6 Supplementary Problems

1. Find all functions $f(x)$ defined for all x, such that

$$|f(x) - f(y)| \leq M(x-y)^2$$

   for all x and y.

2. (*Generalized Rolle's Theorem*) Let $f(x)$ be n times differentiable on $a < x < b$ and assume that $f^{(k)}(x)$ is continuous on $a \leq x \leq b$. Prove that if
$$f(a) = f'(a) = f''(a) = \ldots = f^{(n-1)}(a) = f(b) = 0,$$
then there exists a point c, $a < c < b$, with
$$f^{(n)}(c) = 0.$$

*3. Prove that if $0 \leq x$, y, and $0 < b < a$, then
$$(x^a + y^a)^{1/a} \leq (x^b + y^b)^{1/b}.$$

(*Hint:* First reduce this to the inequality
$$1 + s \leq (1+s)^p$$
for $s > 0$, and $p > 1$.)

4. Let $p(x) = c_0 + c_1 x + c_2 x^2 + \cdots + c_n x^n$ be a polynomial. Prove that if
$$c_0 + \frac{c_1}{2} + \frac{c_2}{3} + \cdots + \frac{c_n}{n+1} = 0,$$
then $p(x)$ has a root between 0 and 1. (*Hint:* Let
$$g(x) = c_0 x + \frac{c_1}{2} x^2 + \cdots + \frac{c_n}{n+1} x^{n+1}.)$$

5. Show that
$$f(x) = \frac{x}{1 + e^{1/x}}$$
$$f(0) = 0$$
defines a continuous function on $\mathbb{R}$. Is $f(x)$ differentiable at $x = 0$?

6. Let $f(x)$ be continuous and differentiable on $[0, 1]$ with
$$f(0) = f(1) = 0.$$
Prove that for every real number a there exists a c, $0 < c < 1$ with $f'(c) = af(c)$.

(*Hint:* Apply Rolle's Theorem to $g(x)f(x)$ for an appropriate function $g(x)$.)

7. Let $f(x)$ be differentiable for $x > 0$ and continuous at $x = 0$, with $f(0) = 0$. If $f'(x)$ is increasing, prove that $g(x) = f(x)/x$ is also increasing.

Chapter 5

Integrals

5.1 The Riemann Integral

The integral of a function $f(x)$ may be thought of as a sum of the values of $f(x)$ over a continuous range of the variable x. To make this notion precise, the integral is defined as the limit of certain finite sums, which geometrically approximate the (signed) area under the graph of $f(x)$. This process, although simple in principle, is somewhat more complicated technically than what has gone before.

The basic idea goes back to the Greeks. It is attributed to Eudoxus, and was used by Archimedes.

Partitions

Let $[a, b]$ be a *finite, closed* interval. A *partition* π of $[a, b]$ is a finite sequence of points

$$a = x_0 < x_1 < \ldots < x_m = b$$

dividing $[a, b]$ into m subintervals $[x_{i-1}, x_i]$. Let

$$\Delta x_i = x_i - x_{i-1}$$

be the length of the ith subinterval. The *norm of* π is the length

$$|\pi| = \max\{\Delta x_1, \ldots, \Delta x_m\}$$

of its longest subinterval.

Definition 1 We say that π' is a *refinement* of π, and write

$$\pi' \supset \pi$$

if π' is obtained by adding points to π.

We denote by

$$\pi = \pi_1 \cup \pi_2$$

the partition π consisting of all the points of π_1 together with all the points of π_2, arranged in increasing order.

Clearly, $\pi_1 \cup \pi_2$ is a *common refinement* of π_1 and π_2; that is, $\pi_1 \cup \pi_2$ is a refinement of both π_1 and π_2. In particular, the partition

$$\pi' = \pi \cup \{c\}$$

formed by adding the single point c to π is a refinement of π. Any refinement of π can be obtained by adding one point at a time to π.

Darboux Sums

For a given partition of $[a, b]$, we shall define the *upper and lower Darboux sums*, which are, respectively, upper and lower approximations to the area under the graph of $f(x)$.

Let $f(x)$ be a *bounded* function on $[a, b]$, and π a partition of $[a, b]$. Define

$$M_i = \sup\{f(x) : x_{i-1} \le x \le x_i\}$$

and

$$m_i = \inf\{f(x) : x_{i-1} \le x \le x_i\}.$$

These numbers are both finite *because $f(x)$ is bounded.*

Define the *upper and lower Darboux sums* by

$$U(f, \pi) = \sum_{i=1}^{n} M_i \Delta x_i$$

and

$$L(f, \pi) = \sum_{i=1}^{n} m_i \Delta x_i.$$

A fixed function f is understood in the following proposition.

Lemma 1

(a) *If $\pi' \supset \pi$, then*
$$L(\pi) \leq L(\pi') \leq U(\pi') \leq U(\pi).$$

(b) *For any two partitions π and π',*
$$L(\pi') \leq U(\pi)$$

(c)
$$-U(-f, \pi) = L(f, \pi).$$

Proof

(a) By the Remark after Definition 1, it suffices to prove the result when $\pi' = \pi \cup \{c\}$ where $x_{k-1} < c < x_k$. In this case, $U(\pi')$ is obtained from $U(\pi)$ by replacing the term
$$M_k \Delta x_k \tag{5.1}$$
by
$$M_k^{(1)}(c - x_{k-1}) + M_k^{(2)}(x_k - c) \tag{5.2}$$
where
$$M_k^{(1)} = \sup\{f(x) : x_{k-1} \leq x \leq c\}$$
and
$$M_k^{(2)} = \sup\{f(x) : c \leq x \leq x_k\}.$$

Since both $M_k^{(1)}$ and $M_k^{(2)}$ are less than or equal to M_k, (5.2) is no larger than (5.1). So $U(\pi') \leq U(\pi)$.

Similarly, $L(\pi) \leq L(\pi')$.

(b) The partitions π and π' have the common refinement $\pi \cup \pi'$. Therefore,
$$L(\pi') \leq L(\pi \cup \pi') \leq U(\pi \cup \pi') \leq U(\pi)$$
by (a).

(c) This follows from the fact that, for any set S, $\sup(-S) = -\inf S$. □

Important Remark In order for this definition to make sense, it is necessary that $f(x)$ be a *bounded* function on a *finite interval* $[a, b]$. For if $f(x)$ is unbounded, at least one of the M_k or m_k must be infinite, while if $[a, b]$ is infinite, at least one of the $\Delta x'_k s$ must be infinite. In either case, at least one of the Darboux sums will not be defined.

The Riemann Integral

We can now define the Riemann integral of $f(x)$. It follows from Lemma 1(b) that

$$\sup_{\pi} L(f, \pi) \leq \inf_{\pi} U(f, \pi).$$

The numbers

$$\bar{I} = \inf_{\pi} U(f, \pi) \quad \text{and} \quad \underline{I} = \sup_{\pi} L(f, \pi)$$

are called the *upper* and *lower integrals* of $f(x)$ over $[a, b]$. Geometrically, $\bar{I}$ is an upper approximation to the (signed) area under the graph of $f(x)$ and the $\underline{I}$ is a lower approximation. If these two approximations are equal, the function $f(x)$ is called *integrable*, and their common value I is called the *integral* of f over $[a, b]$.

Definition 2 *Let $f(x)$ be a bounded function on a finite interval $[a, b]$. We say that $f(x)$ is integrable on $[a, b]$ iff*

$$\sup_{\pi} L(f, \pi) = \inf_{\pi} U(f, \pi).$$

The number

$$I = \sup_{\pi} L(f, \pi) = \inf_{\pi} U(f, \pi)$$

is then called the Riemann integral of f over $[a, b]$, and is denoted by

$$I = \int_a^b f(x)dx.$$

5.1.1 Problems

1. Show directly from the definition that the function

$$f(x) = 1 \qquad a \leq x \leq b$$

is integrable. Find its integral.

2. Show directly from the definition that the function

$$f(x) = \begin{cases} 0 & x \neq c \\ 1 & x = c \end{cases}$$

is integrable on $[a, b]$, where $a < c < b$. Find its integral.

3. Show that the function

$$\chi_{\mathbb{Q}}(x) = \begin{cases} 1 & \text{if } x \text{ is rational} \\ 0 & \text{if } x \text{ is irrational} \end{cases}$$

is *not integrable* on $[0, 1]$.

4. Prove that $f(x) = x$ is integrable on $[0, 1]$. Find its integral. You may wish to use the formula

$$\sum_{k=1}^{n} k = \frac{n(n+1)}{2}.$$

5. Prove that $f(x) = x^2$ is integrable on $[0, 1]$. Find its integral. You may wish to use the formula

$$\sum_{k=1}^{n} k^2 = \frac{n(n+1)(2n+1)}{6}.$$

5.2 Properties of the Integral

The Riemann integral has the following properties.

Theorem 1 (Linearity) *Let $f(x)$ and $g(x)$ be integrable on $[a, b]$.*

(a) *The function $f(x) + g(x)$ is integrable on $[a, b]$, and*

$$\int_a^b f(x) + g(x)\, dx = \int_a^b f(x)\, dx + \int_a^b g(x)\, dx.$$

(b) *If c is a constant, the function $c f(x)$ is integrable on $[a, b]$ and*

$$\int_a^b c f(x)\, dx = c \int_a^b f(x)\, dx.$$

Lemma 2 *For any bounded functions f and g and any partition π of $[a,b]$, we have*

$$U(f+g, \pi) \leq U(f, \pi) + U(g, \pi).$$

Proof of Lemma 2 We have

$$\begin{aligned} M_i(f+g) &= \sup\{f(x) + g(x) : x_{i-1} \leq x \leq x_i\} \\ &\leq \sup\{f(x) : x_{i-1} \leq x \leq x_i\} + \sup\{g(x) : x_{i-1} \leq x \leq x_i\} \\ &= M_i(f) + M_i(g). \end{aligned}$$

Hence,

$$\begin{aligned} U(f+g, \pi) &= \sum_{i=1}^{n} M_i(f+g) \Delta x_i \\ &\leq \sum_{i=1}^{n} M_i(f) \Delta x_i + \sum_{i=1}^{n} M_i(g) \Delta x_i \\ &= U(f, \pi) + U(g, \pi). \end{aligned} \qquad \square$$

Proof of Theorem 1 Let $\epsilon > 0$. Since, by definition,

$$I(f) = \int_a^b f(x)dx = \inf_\pi U(f, \pi),$$

we can choose π_1 such that

$$U(f, \pi_1) < I(f) + \frac{\epsilon}{2}$$

and similarly, π_2 such that

$$U(g, \pi_2) < I(g) + \frac{\epsilon}{2}.$$

Letting $\pi = \pi_1 \cup \pi_2$, we have, by Lemma 2,

$$\begin{aligned} \bar{I}(f+g) &\leq U(f+g, \pi) \leq U(f, \pi) + U(g, \pi) \\ &\leq U(f, \pi_1) + U(g, \pi_2) \leq I(f) + I(g) + \epsilon. \end{aligned}$$

Since ϵ is arbitrary, we conclude that

$$\bar{I}(f+g) \leq I(f) + I(g).$$

By a similar argument with lower sums, we also obtain

$$\underline{I}(f+g) \geq \underline{I}(f) + \underline{I}(g).$$

Thus,

$$I(f) + I(g) \geq \bar{I}(f+g) \geq \underline{I}(f+g) \geq I(f) + I(g),$$

which implies that f is integrable, and $I(f+g) = I(f) + I(g)$.
(b) See Problem 2. □

Theorem 2 (Positivity) *Let $f(x)$ be integrable on $[a,b]$ and $f(x) \geq 0$. Then*

$$\int_a^b f(x)dx \geq 0.$$

Proof Clearly, $m_i \geq 0$, so that for any π

$$\int_a^b f(x)dx \geq L(f,\pi) \geq 0. \qquad \square$$

Theorem 3 *Let $f(x)$ be a bounded function on $[a,b]$ and $a < c < b$. If $f(x)$ is integrable on $[a,c]$ and $[c,b]$, then $f(x)$ is also integrable on $[a,b]$ and*

$$\int_a^b f(x)dx = \int_a^c f(x)dx + \int_c^b f(x)dx.$$

Proof Let $\epsilon > 0$. Let

$$I_1 = \int_a^c f(x)dx \quad \text{and} \quad I_2 = \int_c^b f(x)dx.$$

As above, choose a partition π_1 of $[a,c]$ such that

$$U(f,\pi_1) < I_1 + \frac{\epsilon}{2}$$

and a partition π_2 of $[c,b]$ such that

$$U(f,\pi_2) < I_2 + \frac{\epsilon}{2}.$$

Let $\pi = \pi_1 \cup \pi_2$ be the partition of $[a,b]$ formed by taking all points of both partitions π_1 and π_2. If

$$\bar{I}(f) = \inf_\pi U(f,\pi),$$

then
$$\bar{I}(f) \leq U(f,\pi) = U(f,\pi_1) + U(f,\pi_2) < I_1 + I_2 + \epsilon.$$

Since ϵ is arbitrary, we conclude that
$$\bar{I}(f) \leq I_1 + I_2.$$

By a similar argument with lower sums, we also obtain
$$\underline{I}(f) \geq I_1 + I_2.$$

As above, we therefore obtain that f is integrable, and $I(f) = I_1 + I_2$. □

5.2.1 Problems

1. (a) Prove that if $f(x)$ is integrable on $[a,b]$ and $|f(x)| \leq M$, then
$$\left| \int_a^b f(x)\, dx \right| \leq M(b-a).$$

   (b) Assume that $|f(x)|$ is integrable, and prove that
$$\left| \int_a^b f(x)\, dx \right| \leq \int_a^b |f(x)|\, dx.$$

   (*Remark:* It is proved in Problem 1 of Section 5.3, that if $f(x)$ is integrable, then so is $|f(x)|$.)

2. Prove (b) of Theorem 1. (*Hint:* You may need to separate cases according to whether c is positive or negative.)

3. Show that if $f(x)$ is integrable on $[a,b]$, and if $g(x)$ is equal to $f(x)$ except at a finite number of points, then $g(x)$ is also integrable and their integrals are equal. (*Hint:* Use Problem 2 of Section 5.1.)

5.3 Riemann's Integrability Condition

The following is a very useful, necessary, and sufficient condition for a function to be integrable.

Theorem 4 (Riemann's Integrability Condition) *The bounded function $f(x)$ is integrable on $[a,b]$ iff for every $\epsilon > 0$, there is a partition π such that*
$$U(f,\pi) - L(f,\pi) < \epsilon. \tag{5.3}$$

Proof For any π, we have
$$L(f,\pi) \leq \underline{I} \leq \overline{I} \leq U(f,\pi).$$

If (5.3) holds, then
$$0 \leq \overline{I} - \underline{I} \leq U(f,\pi) - L(f,\pi) < \epsilon$$

for every $\epsilon > 0$. Hence, $\overline{I} = \underline{I}$.

Conversely, there exists a π_1 such that
$$U(f,\pi_1) - \overline{I} < \frac{\epsilon}{2}$$

and a π_2 such that
$$\underline{I} - L(f,\pi_2) < \frac{\epsilon}{2}.$$

If $f(x)$ is integrable, then $\overline{I} = \underline{I} = I$, so if $\pi = \pi_1 \vee \pi_2$, then
$$\begin{aligned}U(f,\pi) - L(f,\pi) &= (U(f,\pi) - I) + (I - L(f,\pi))\\ &\leq (U(f,\pi_1) - I) + I - L(f,\pi_2)\\ &< \frac{\epsilon}{2} + \frac{\epsilon}{2} = \epsilon.\end{aligned}$$
□

Corollary 1 *If $f(x)$ is integrable on $[a,b]$, then so is $|f(x)|$.*

The proof is sketched in Problem 1 below.

5.3.1 Problems

1. Let $f(x)$ be a function on $[a,b]$. Define the positive and negative parts of $f(x)$ to be
$$f_+(x) = \max\{f(x), 0\}$$

and
$$f_-(x) = -\max\{-f(x), 0\} = -\min\{f(x), 0\};$$

or, in other words,
$$f_+(x) = \begin{cases} f(x) & \text{if } f(x) \geq 0 \\ 0 & \text{if } f(x) < 0 \end{cases}$$
$$f_-(x) = \begin{cases} 0 & \text{if } f(x) \geq 0 \\ -f(x) & \text{if } f(x) < 0. \end{cases}$$

(a) Prove that $f_\pm(x) \geq 0$,
$$f(x) = f_+(x) - f_-(x)$$
and
$$|f(x)| = f_+(x) + f_-(x).$$

(b) Prove that if $f(x)$ is integrable, then $f_+(x)$ is integrable. (*Hint:* Note that $M_i^+ = \max[M_i, 0]$ and $m_i^+ = \max[m_i, 0]$ so that $M_i^+ - m_i^+ \leq M_i - m_i$.)

(c) Use Parts (a) and (b) to show that if $f(x)$ is integrable, then so are $f_-(x)$ and $|f(x)|$.

2. Prove that the function defined by $f(0) = 0$ and
$$f(x) = \frac{1}{2^n} \quad \text{if} \quad \frac{1}{2^n} < x \leq \frac{1}{2^{n-1}}$$
for $n = 0, 1, 2, \ldots$ is integrable on $[0, 1]$.

5.4 Integrability Theorems

We now shall prove several theorems concerning which functions are integrable. All the proofs are based on the Riemann Condition.

Theorem 5 *If $f(x)$ is integrable on $[a, b]$ and $a \leq c \leq d \leq b$, then $f(x)$ is integrable on $[c, d]$.*

Proof Let $\epsilon > 0$. Choose π with
$$U(\pi) - L(\pi) < \epsilon.$$

By Lemma 1(b), we may assume that c and d are points of π, since adding points to π can only decrease the difference. If we omit the terms in the sum
$$U(\pi) - L(\pi) = \sum_{i=1}^{n}(M_i - m_i)\Delta x_i$$
corresponding to subintervals outside of $[c, d]$, then we get the difference of the upper and lower sums for a partition π' of $[c, d]$. The omitted terms are all positive since $(M_i - m_i) \geq 0$, so we have
$$U(\pi') - L(\pi') \leq U(\pi) - L(\pi) < \epsilon. \qquad \square$$

Theorem 6 *If $f(x)$ is monotone on $[a, b]$, then $f(x)$ is integrable on $[a, b]$.*

Proof Let $f(x)$ be increasing on $[a,b]$. If π is any partition of $[a,b]$, then
$$M_i = f(x_i) \quad \text{and} \quad m_i = f(x_{i-1}).$$

Choose π to be a partition with n intervals of equal length
$$\Delta x = \frac{(b-a)}{n}.$$

Then
$$U(\pi) - L(\pi) = \sum_{i=1}^{n}(M_i - m_i)\Delta x_i$$
$$= \frac{(b-a)}{n}\sum_{i=1}^{n}(f(x_i) - f(x_{i-1}))$$
$$= \frac{(b-a)}{n}[f(b) - f(a)].$$

This will be less than ϵ if
$$n > \frac{(b-a)}{\epsilon}[f(b) - f(a)]. \qquad \square$$

Theorem 7 *If $f(x)$ is integrable on $[a,b]$ and $f(x) \geq 0$, then $f(x)^2$ is also integrable on $[a,b]$.*

Proof Let $0 \leq f(x) \leq M$. Choose π such that
$$U(f,\pi) - L(f,\pi) < \frac{\epsilon}{2M}.$$

Clearly,
$$0 \leq m_i \leq M_i \leq M.$$

Since $f(x)$ is positive, we also have
$$M_i(f^2) = M_i(f)^2$$
and
$$m_i(f^2) = m_i(f)^2.$$

Thus

$$U(f^2, \pi) - L(f^2, \pi) = \sum_{i=1}^{n} \left[M_i(f^2) - m_i(f^2)\right] \Delta x_i$$

$$= \sum_{i=1}^{n} \left[M_i(f)^2 - m_i(f)^2\right] \Delta x_i$$

$$= \sum_{i=1}^{n} \left[M_i(f) + m_i(f)\right]\left[M_i(f) - m_i(f)\right] \Delta x_i$$

$$\leq 2M \sum_{i=1}^{n} \left[M_i(f) - m_i(f)\right] \Delta x_i$$

$$= 2M \left[U(f, \pi) - L(f, \pi)\right] < 2M \frac{\epsilon}{2M} = \epsilon. \qquad \square$$

Corollary 2 *If $f(x)$ and $g(x)$ are integrable on $[a, b]$, then the product $f(x)g(x)$ is also integrable on $[a, b]$.*

Proof Let f be integrable on $[a, b]$, and $|f(x)| \leq M$. Then $f + M \geq 0$, so by Theorem 7, $(f + M)^2$ is integrable. Hence, so is

$$f^2 = (f + M)^2 - 2Mf - M^2.$$

Thus, the square of an arbitrary integrable function is integrable.

It follows that if f and g are integrable, then so is

$$fg = \frac{1}{4}\left\{(f + g)^2 - (f - g)^2\right\}$$

since $f + g$ and $f - g$ are integrable functions. $\qquad \square$

5.4.1 Problems

1. Prove that the function

$$\psi(x) = \begin{cases} 1/q & \text{if } x = p/q \text{ in lowest terms} \\ 0 & \text{if } x \text{ is irrational} \end{cases}$$

is integrable on $[0, 1]$ and find its integral.

2. Let $\psi(x)$ be as defined in Problem 1, and

$$H(x) = \begin{cases} 1 & \text{if } x > 0 \\ 0 & \text{if } x \leq 0. \end{cases}$$

Show that $H(\psi(x))$ is not integrable. Thus, *the composition of integrable functions need not be integrable.*

5.5 Uniform Continuity

We now introduce the rather technical notion of *uniform continuity*. This subject would have been more logically taken up in Chapter 3, since it has nothing to do with integrals. However, it plays an essential role in the proof of the next section that continuous functions are integrable, so we have chosen to discuss it here where we can provide an instance of its use.

Definition 3 A function $f(x)$ on a set D is *uniformly continuous on D* iff for every positive number ϵ there is a positive number δ, such that

$$|f(x) - f(y)| < \epsilon$$

whenever $x, y \in D$ and $|x - y| < \delta$.

Remark This looks very much like the ϵ, δ-definition of continuity. The difference is that *the same δ works for all points x and y*. This is unpleasantly technical, but is nevertheless important, and cannot be avoided.

We will illustrate the notion with a few examples.

Example 1 Let $f(x) = x^2$ and $D = [0, 1]$. The function f *is uniformly continuous on D*.

For we have

$$|f(x) - f(y)| = |x^2 - y^2| = |x - y|\,|x + y| \leq 2\,|x - y|.$$

Hence, if

$$\delta = \frac{\epsilon}{2},$$

we have

$$|f(x) - f(y)| \leq 2\delta < 2 \cdot \frac{\epsilon}{2} = \epsilon$$

whenever $|x - y| < \delta$. ∎

Example 2 Let $f(x) = x^2$ and $D = [0, \infty)$. The function f is *not uniformly continuous on D*. For we have

$$|f(x) - f(y)| = |x^2 - y^2| = |x - y|\,|x + y|$$

so that

$$|f(x) - f(y)| \geq 2\,|x - y|\min(x, y).$$

Therefore, if we want to make $|f(x) - f(y)|$ less than ϵ, we have to make $|x - y|$ less than

$$\frac{\epsilon}{2\min(x,y)}.$$

But this number gets smaller and smaller as x and y get large. So no positive number δ can work for *all* x and y.

This can be seen from the graph of $f(x) = x^2$. As the two points x and y get farther to the right, the function values $f(x)$ and $f(y)$ differ by larger and larger amounts for the same fixed distance between x and y. ∎

Note from these examples that uniform continuity depends not only on $f(x)$, but *on the domain D as well.*

Example 3 Let $f(x) = 1/x$ and $D = (0, 1]$. The function f is *not uniformly continuous on D.*

For, again from the graph, the function varies rapidly over small intervals when x and y are very small. Analytically, we have

$$|f(x) - f(y)| = \left|\frac{1}{x} - \frac{1}{y}\right| = \frac{|x-y|}{xy}.$$

If this is to be made less than ϵ, then $|x - y|$ must be less than ϵxy. This quantity is smaller and smaller for x and y close to zero, so no single number δ can suffice for all x and y. ∎

The important result on uniform continuity is the following.

Theorem 8 *If $f(x)$ is continuous on a closed, bounded interval $[a, b]$, then $f(x)$ is uniformly continuous on $[a, b]$.*

Proof Suppose that $f(x)$ is *not* uniformly continuous. Then there is *some* $\epsilon > 0$, such that for *every* $\delta > 0$, there are points x and y for which

$$|f(x) - f(y)| \geq \epsilon.$$

If we fix this ϵ and choose $\delta = 1/n$, then there must exist points x_n and y_n such that

$$|f(x_n) - f(y_n)| \geq \epsilon \quad \text{and} \quad |x_n - y_n| < \frac{1}{n}. \tag{5.4}$$

By the Bolzano–Weierstrass Theorem, x_n has a subsequence x_{n_k} convergent to a point c of $[a, b]$:

$$x_{n_k} \to c.$$

Moreover,
$$y_{n_k} = x_{n_k} + (y_{n_k} - x_{n_k}) \to c + 0 = c$$
as well.

Passing to the limit in (5.4), we find that since f is continuous at c,
$$0 = |f(c) - f(c)| = \lim |f(x_{n_k}) - f(y_{n_k})| \geq \epsilon > 0,$$
which is a contradiction. Therefore, $f(x)$ is uniformly continuous. □

5.5.1 Problems

1. Where in the proof of Theorem 8 did we use (a) continuity and (b) that $[a, b]$ is closed and bounded?

2. (*Sufficient Condition for Uniform Continuity*) Let $f(x)$ be differentiable on $a < x < b$, and assume that $f'(x)$ is bounded. Prove that $f(x)$ is uniformly continuous on $a < x < b$.

3. Is the converse of Problem 2 true? Justify your answer with either a proof or a counterexample.

4. Are the following functions uniformly continuous on the given set?

    (a) $x^2 + 2x + 3$ on $[0, 1]$
    (b) $\sin x$ on $\mathbb{R}$
    (c) $\frac{1}{\sqrt{x}}$ on $[1, \infty)$
    (d) $\frac{1}{\sqrt{x}}$ on $(0, 1)$
    (e) $\sqrt{x}$ on $[0, \infty)$
    (f) $\frac{x^2}{x+1}$ on $[0, \infty)$

5. Is the function
$$f(x) = |x - a| - |x - b|$$
uniformly continuous on $\mathbb{R}$?

*6. Let S be a subset of the real numbers, and x a point. Define the *distance from x to S* to be
$$d(x, S) = \inf \{|x - s| : s \in S\}.$$
Prove that $d(x, S)$ is uniformly continuous on $\mathbb{R}$.
(*Hint:* Prove that $|d(x, S) - d(y, S)| \leq |x - y|$.)

7. If f is uniformly continuous on a set S, and a_n is a Cauchy sequence contained in S, then $f(a_n)$ is a Cauchy sequence. Is this true if f is continuous on S, but not *uniformly* continuous?

8. Prove that if f is uniformly continuous on a bounded set S, then f is bounded on S.

5.6 Integrability of Continuous Functions

We can now prove that continuous functions are integrable.

Theorem 9 *If $f(x)$ is continuous on $[a,b]$, then $f(x)$ is integrable on $[a,b]$.*

Proof Let $\epsilon > 0$. By uniform continuity, we can choose $\delta > 0$ such that
$$|f(x) - f(y)| < \frac{\epsilon}{2(b-a)}$$
if $|x - y| < \delta$. If π is a partition with $|\pi| < \delta$, then
$$M_k - m_k \leq \frac{\epsilon}{2(b-a)}.$$
Hence,
$$U(\pi) - L(\pi) = \sum_{k=1}^{n} (M_k - m_k) \Delta x_k$$
$$\leq \frac{\epsilon}{2(b-a)} \sum_{k=1}^{n} \Delta x_k = \frac{\epsilon}{2(b-a)} \cdot (b-a) = \frac{\epsilon}{2} < \epsilon.$$

Thus, by the Riemann Condition, $f(x)$ is integrable. □

5.6.1 Problems

1. Let $f(x)$ be continuous, and $f(x) \geq 0$ on $[a,b]$. Prove that if
$$\int_a^b f(x)dx = 0,$$
then $f(x) = 0$ for all x.

2. Let $f(x)$ be continuous on $[a,b]$ and assume that
$$\int_a^b f(x)dx = 0.$$
Prove that there exists a point c in $[a,b]$ where $f(c) = 0$.

3. Let $p(x) = c_0 + c_1 x + c_2 x^2 + \cdots + c_n x^n$ be a polynomial. Prove that if
$$c_0 + \frac{c_1}{2} + \frac{c_2}{3} + \cdots + \frac{c_n}{n+1} = 0,$$
then $p(x)$ has a root between 0 and 1. (*Hint:* This was a problem in Chapter 4. This time, use Integrals.)

5.7 *Riemann Sums

Let $f(x)$ be a bounded function on $[a, b]$, and π a partition of $[a, b]$. Choose a point ξ_i in the ith subinterval:
$$x_{i-1} \leq \xi_i \leq x_i,$$
and let $\xi = (\xi_1, \ldots, \xi_m)$. Define the *Riemann sum*
$$R(f, \pi, \xi) = \sum_{i=1}^{m} f(\xi_i) \Delta x_i.$$

The following result holds regardless of how the vectors $\xi(n)$ are chosen.

Theorem 10 *Let $f(x)$ be integrable on $[a, b]$. Let π_n be a sequence of partitions with $|\pi_n| \to 0$, and let $\xi(n) = (\xi_1(n), \ldots, \xi_m(n))$ be any choice of the points $\xi_i(n)$ in $[x_{i-1}, x_i]$. Then*
$$\lim R(f, \pi_n, \xi(n)) = \int_a^b f(x)\,dx. \tag{5.5}$$

We express (5.5) by writing
$$\lim_{|\pi| \to 0} R(f, \pi, \xi) = \int_a^b f(x)\,dx.$$

We shall require two Lemmas.

Lemma 3 *Let $|f(x)| \leq M$. Let π be a partition of $[a, b]$, and let $\pi' = \pi \cup \{c\}$ be the partition obtained from π by adding the point c. Then*
$$|U(\pi) - U(\pi')| \leq M\,|\pi|.$$

Proof Let $x_{i-1} < c < x_i$. Then $U(\pi')$ is obtained from $U(\pi)$ by replacing
$$M_i \Delta x_i = M_i(x_i - x_{i-1})$$

by
$$M'_i (c - x_{i-1}) + M''_i (x_i - c).$$
Since at least one of M'_i and M''_i is equal to M_i, the difference is either
$$(M_i - M'_i)(c - x_{i-1})$$
or
$$(M_i - M''_i)(x_i - c),$$
neither of which exceeds $M |\pi|$. □

Lemma 4 *Assume that $f(x)$ is integrable. If $|\pi_n| \to 0$, then*
$$U(\pi_n) - L(\pi_n) \to 0 \tag{5.6}$$
and hence
$$\lim U(\pi_n) = \lim L(\pi_n) = I. \tag{5.7}$$

Proof Let $\epsilon > 0$, and choose π_0 so that
$$U(\pi_0) - L(\pi_0) < \frac{\epsilon}{2}.$$
If N is the number of points in π_0, then applying Lemma 3 N times yields
$$|U(\pi_0 \vee \pi_n) - U(\pi_n)| \leq MN |\pi_n|$$
and
$$|L(\pi_0 \vee \pi_n) - L(\pi_n)| \leq MN |\pi_n|.$$
Thus,
$$\begin{aligned} U(\pi_n) - L(\pi_n) &\leq \{U(\pi_0 \vee \pi_n) + MN |\pi_n|\} - \{L(\pi_0 \vee \pi_n) - MN |\pi_n|\} \\ &= U(\pi_0 \vee \pi_n) - L(\pi_0 \vee \pi_n) + 2MN |\pi_n| \\ &\leq U(\pi_0) - L(\pi_0) + 2MN |\pi_n| \\ &< \frac{\epsilon}{2} + 2MN |\pi_n| < \frac{\epsilon}{2} + \frac{\epsilon}{2} = \epsilon \end{aligned}$$
provided we choose n so large that
$$|\pi_n| < \frac{\epsilon}{4MN}.$$
□

Proof of Theorem 10 We have
$$L(\pi_n) \leq R(\xi, \pi_n) \leq U(\pi_n).$$
Hence,
$$\lim R(\xi, \pi_n) = I$$
by the Sandwich Theorem. □

Example Partition the interval $[0,1]$ into n equal intervals, and take $f(x) = x$. Then
$$\lim \sum_{k=1}^{n} \left(\frac{k}{n}\right)\frac{1}{n} = \lim \frac{1}{n^2}\sum_{k=1}^{n} k = \int_0^1 x\,dx = \frac{1}{2}.$$ ■

5.7.1 Problems

1. Show that for $f(x)$ continuous
$$\int_0^1 f(x)\,dx = \lim_{n\to\infty} \sum_{k=1}^{n} f\left(\frac{k}{n}\right)\frac{1}{n}.$$

For Problems 2, 3, and 4, calculate the resulting integrals as usual with the Fundamental Theorem of Calculus, which is proved in the next section.

2. Evaluate
$$\lim_{n\to\infty} \sum_{k=1}^{n} \frac{k^2}{n^3}.$$

3. Evaluate
$$\lim_{n\to\infty} \frac{1}{n^{p+1}} \sum_{k=1}^{n} k^p$$
for $p > 0$.

4. Prove that the sequence
$$a_n = \frac{1}{n+1} + \frac{1}{n+2} + \ldots + \frac{1}{2n}$$
converges to $\log 2$.

5. Prove that (5.6) implies (5.7).

*6. (*Alternate definition of Integrability*) Let $f(x)$ be bounded on $[a,b]$. Prove that $f(x)$ is integrable on $[a,b]$ iff there is a number I such that for every $\epsilon > 0$, there exists a $\delta > 0$ such that
$$|R(\xi, \pi) - I| < \epsilon$$
whenever $|\pi| < \delta$, regardless of how ξ is chosen.

Prove that when this is so,
$$I = \int_a^b f(x)\,dx.$$

5.8 The Fundamental Theorem

The Fundamental Theorem of Calculus states that Differentiation and Integration are inverse operations. We shall prove two versions of this Theorem by two separate methods. The first is the usual version.

Theorem 11 (Fundamental Theorem of Calculus I) *Let $f(x)$ be continuous on $[a,b]$. Define*
$$F(x) = \int_a^x f(t)\,dt$$
for $a \le x \le b$. If $a < c < b$, then
$$F'(c) = f(c).$$

Proof The difference quotient at c is
$$\frac{F(c+h) - F(c)}{h} = \frac{1}{h}\int_c^{c+h} f(t)\,dt.$$

Hence,
$$\frac{F(c+h) - F(c)}{h} - f(c) = \frac{1}{h}\int_c^{c+h} f(t) - f(c)\,dt.$$

Let $\epsilon > 0$. By continuity of f at c, we may choose a $\delta > 0$ such that
$$|f(c) - f(t)| < \epsilon$$
whenever $|c - t| < \delta$. Hence, if $|h| < \delta$, we have
$$\left|\frac{F(c+h) - F(c)}{h} - f(c)\right| \le \frac{1}{|h|}\left|\int_c^{c+h} |f(t) - f(c)|\,dt\right|$$
$$\le \frac{1}{|h|}\left|\int_c^{c+h} \epsilon\,dt\right| = \frac{1}{|h|}|h|\epsilon = \epsilon. \qquad \square$$

5.8 The Fundamental Theorem

Corollary 3 Let $f(x)$ be continuous and $F'(x) = f(x)$. Then
$$\int_a^b f(x)dx = F(b) - F(a).$$

Proof If we define
$$g(x) = \int_a^x f(t)dt - F(x),$$
then by Theorem 11,
$$g'(x) = f(x) - f(x) = 0$$
for all x, so that
$$g(x) = C$$
where C is a constant. But
$$C = g(a) = 0 - F(a) = -F(a)$$
so
$$g(x) = F(x) - F(a).$$
Set $x = b$ to obtain the result. $\square$

The second version is somewhat more general, and uses Riemann sums in the proof.

Theorem 12 (Fundamental Theorem of Calculus II) Let $f(x)$ be differentiable and $f'(x)$ be integrable on $[a,b]$. Then
$$\int_a^b f'(x)dx = f(b) - f(a). \tag{5.8}$$

Proof We have
$$f(b) - f(a) = \sum_{k=1}^n \left[f(x_i) - f(x_{i-1}) \right].$$
By the Mean Value Theorem, there is a number c_i in $[x_{i-1}, x_i]$ with
$$f(x_i) - f(x_{i-1}) = f'(c_i).$$
Thus
$$f(b) - f(a) = \sum_{k=1}^n f'(c_i)\Delta x_i.$$

The right side is a Riemann sum, and so tends to

$$\int_a^b f'(x)dx$$

as $|\pi| \to 0$. Since the left side is independent of π, we obtain (5.8). □

5.8.1 Problems

1. Find the derivatives of

   (a) $F(x) = \int_0^x e^{t^2} dt$, and
   (b) $G(x) = \int_{x^2}^x (\sin t)/t \, dt$.

2. By the Fundamental Theorem, we have

   $$\int_{-1}^1 \frac{1}{x^2} dx = \left[-\frac{1}{x}\right]_{-1}^1 = -2.$$

   Why is the answer negative when the integrand is positive?

3. (*Mean Value Theorem for Integrals*) Let $f(x)$ and $g(x)$ be continuous on $[a,b]$, and $g(x) > 0$. Prove that there exists a point c in $[a,b]$ such that

   $$\int_a^b f(x)g(x) \, dx = f(c) \int_a^b g(x) \, dx.$$

   (*Hint:* There is a proof using the Intermediate Value Theorem, and another using the Fundamental Theorem and Cauchy's Mean Value Theorem. Find both.)

   What happens if $g(x)$ changes sign? Give an example.

4. Give an example of an integral to which Theorem 12 applies, but Theorem 11 does not.

5.9 Substitution and Integration by Parts

We shall prove two important consequences of the Fundamental Theorem.

Recall that we say that a function $f(x)$ is *continuously differentiable* iff $f'(x)$ exists and is continuous.

Theorem 13 (Integration by Substitution) *Let $f(x)$ be continuous on $[a,b]$. Let $x(t)$ be continuously differentiable on $[\alpha, \beta]$ with $a \leq x(t) \leq b$, and*

$$x(\alpha) = a \quad \text{and} \quad x(\beta) = b.$$

5.9 Substitution and Integration by Parts

Then

$$\int_a^b f(x)\,dx = \int_\alpha^\beta f(x(t))x'(t)\,dt. \tag{5.9}$$

Proof Let

$$F(t) = \int_a^x f(s)\,ds.$$

The composition

$$G(t) = F(x(t))$$

is then well-defined. By the Chain Rule and Theorem 11,

$$G'(t) = F'(x(t))x'(t) = f(x(t))x'(t).$$

Hence, by Corollary 3,

$$\int_\alpha^\beta f(x(t))x'(t)\,dt = \int_\alpha^\beta G'(t)\,dt = G(\beta) - G(\alpha)$$
$$= F(b) - F(a) = \int_a^b f(x)\,dx. \qquad \square$$

Remark Note that when changing variables from x to t in a *definite integral* (i.e., one with limits), it is *necessary to change the limits* on x to the limits on t. For this reason, it is sometimes useful to think of (5.9) as

$$\int_{x=a}^{x=b} f(x)\,dx = \int_{t=\alpha}^{t=\beta} f(x(t))x'(t)\,dt.$$

For example, if $x(t) = \cos t$, one has

$$\int_{x=0}^{x=1} e^x\,dx = \int_{t=0}^{t=\pi/2} e^{\sin t} \cos t\,dt.$$

Theorem 14 (Integration by Parts) *Let $u(x)$ and $v(x)$ be continuously differentiable on $[a,b]$. Then*

$$\int_a^b u(x)v'(x)\,dx = [u(x)v(x)]_a^b - \int_a^b u'(x)v(x)\,dx.$$

Proof By the Product Rule,

$$\frac{d}{dx}[u(x)v(x)] = u(x)v'(x) + u'(x)v(x).$$

The result follows by Corollary 3. $\qquad \square$

5.9.1 Problems

1. Use $\cos(\frac{\pi}{2} - x) = \sin x$, to show that

$$\int_0^{\pi/2} \cos^2 x \, dx = \int_0^{\pi/2} \sin^2 x \, dx$$

and then use this to show that

$$\int_0^{\pi/2} \cos^2 x \, dx = \frac{\pi}{4}.$$

2. Let $f(x)$ be continuous on $[a, b]$.

   (a) If $f(-x) = f(x)$, prove that

   $$\int_{-a}^{a} f(x) dx = 2 \int_0^a f(x) dx.$$

   (b) If $f(-x) = -f(x)$, prove that

   $$\int_{-a}^{a} f(x) dx = 0.$$

3. If $f(x)$ is continuous on $\mathbb{R}$, and $f(x+p) = f(x)$, prove that

$$\int_a^b f(x) dx = \int_{a+p}^{b+p} f(x) dx.$$

4. Use the substitution $u = \pi - x$ to show that

$$\int_0^\pi x f(\sin x) \, dx = \frac{\pi}{2} \int_0^\pi f(\sin x) \, dx.$$

Use this fact to evaluate the integral

$$\int_0^\pi \frac{x \sin x}{1 + \cos^2 x} dx.$$

5. Find the result of making the substitution $x = \sin \phi$ in the integral

$$\int_0^1 \frac{1}{\sqrt{(1-x^2)(4-x^2)}} dx.$$

6. Use integration by parts to express
$$\int_0^{\pi/2} \frac{\sin x \cos x}{x+1} dx$$
in terms of
$$I = \int_0^{\pi} \frac{\cos x}{(2+x)^2} dx.$$

7. Prove that if $u(x)$ is twice differentiable on $[0,1]$, and
$$u(0) = u(1) = 0,$$
then
$$\int_0^1 u''(x) u(x)\, dx \leq 0.$$

8. Let
$$I_n = \int_0^1 (1-x^2)^n\, dx.$$

(a) Show that
$$I_n = \frac{2n}{2n+1} I_{n-1}.$$

(b) Use this to find $I_2, I_3, I_4,$ and I_5.

(c) Find a general formula for I_n.

9. Let
$$I_n = \int_0^{\pi/4} \tan^n x\, dx.$$

(a) Show that
$$I_n = \frac{1}{n-1} - I_{n-2}.$$

(b) Compute I_1 and I_2, and use this formula to find $I_3, I_4,$ and I_5.

*(c) Find a general formula for I_n.

*(d) Show that $\lim I_n = 0$. What infinite series have you evaluated? (*Hint:* Fix $\epsilon > 0$, and split the integral into an integral over $\left[0, \frac{\pi}{4} - \epsilon\right]$ and one over $\left[\frac{\pi}{4} - \epsilon, \frac{\pi}{4}\right]$.)

10. If $u(x)$ is twice differentiable on $[0, \pi]$, and $u(0) = u(\pi) = 0$, find
$$\int_0^{\pi} (u''(x) + u(x)) \sin x\, dx.$$

5.10 *Improper Integrals

As we have explained in Section 5.1, the definition of the Riemann integral requires that $f(x)$ be a *bounded* function on a *finite* interval. This requirement is essential for this definition to work. Nevertheless, it is both possible and desirable to integrate unbounded functions and to integrate over infinite intervals. This is done by integrating over a subinterval and passing to the limit.

For example, consider

$$\int_0^1 \frac{1}{\sqrt{x}} dx.$$

The function $1/\sqrt{x}$ is not bounded on the interval of integration $[0, 1]$. However, if we integrate from t to 1 for $t > 0$, and pass to the limit, we get

$$\lim_{t \to 0+} \int_t^1 \frac{1}{\sqrt{x}} dx = \lim_{t \to 0+} \left[2\sqrt{x}\right]_t^1 = \lim_{t \to 0+} \left(2 - 2\sqrt{t}\right) = 2.$$

So we take

$$\int_0^1 \frac{1}{\sqrt{x}} dx = 2.$$

In general, we make the following definition:

We say that the function $f(x)$ defined on a finite or infinite open interval (a, b) is *locally Riemann integrable on* (a, b) iff $f(x)$ is Riemann integrable on every closed subinterval $[c, d] \subset (a, b)$.

Definition 4 If $f(x)$ is locally Riemann integrable on (a, b), the *improper integral*

$$\int_a^b f(x) dx$$

is defined to be

$$\int_a^b f(x) dx = \lim_{s \to a+} \int_s^c f(x) dx + \lim_{t \to b-} \int_c^t f(x) dx$$

where c is any point between a and b.

Definition 5 If $f(x)$ is locally Riemann integrable on (a, b), we say that the *improper integral*

$$\int_a^b f(x) dx$$

is *absolutely convergent* iff the improper integral

$$\int_a^b |f(x)|\,dx$$

exists. Geometrically, this means that the area under the curve $|f(x)|$ is finite.

Theorem 15 *The integral $\int_a^b f(x)dx$ is absolutely convergent iff there exists an M such that*

$$\int_c^d |f(x)|\,dx \le M$$

for every $[c,d] \subset (a,b)$.

Proof The function

$$F(t) = \int_c^t f(x)dx$$

is increasing and bounded. Therefore

$$\lim_{t \to b-} F(t) = \lim_{t \to b-} \int_c^t f(x)dx$$

exists. Similarly for the endpoint a. $\square$

Corollary 4 *If $|f(x)| \le g(x)$ and $\int_a^b g(x)dx$ is absolutely convergent, then $\int_a^b f(x)dx$ is absolutely convergent.*

Theorem 16 *If the integral $\int_a^b f(x)dx$ is absolutely convergent, then it is convergent.*

Proof Let $F(t) = \int_c^t f(x)dx$ and $G(t) = \int_c^t g(x)dx$. Then

$$|F(s) - F(t)| = \left|\int_t^s f(x)dx\right| \le \int_t^s |f(x)|\,dx = G(s) - G(t) \to 0$$

as $t, s \to b-$. Thus, $\lim_{t \to b-} F(t)$ exists by Cauchy's criterion. $\square$

Example 1 Consider $\int_1^\infty \sin x\, e^{-x^2} dx$. We have

$$\left|\sin x\, e^{-x^2}\right| \le e^{-x^2} \le e^{-x}.$$

But
$$\lim_{t\to\infty} \int_1^t e^{-x}\,dx = \lim_{t\to\infty}\left(e^{-1} - e^{-t}\right) = \frac{1}{e}$$
is finite, so the integral
$$\int_1^\infty \sin x\, e^{-x^2}\,dx$$
is *absolutely convergent*. ∎

Example 2 A more complicated example is
$$\int_0^\infty \frac{\sin x}{x}\,dx.$$
This integral is *not absolutely convergent*, since
$$\int_0^{n\pi} \left|\frac{\sin x}{x}\right|\,dx \geq \int_\pi^{n\pi} \left|\frac{\sin x}{x}\right|\,dx = \sum_{k=1}^{n-1} \int_{k\pi}^{(k+1)\pi} \left|\frac{\sin x}{x}\right|\,dx$$
$$\geq \sum_{k=1}^{n-1} \frac{1}{k\pi} \int_{k\pi}^{(k+1)\pi} |\sin x|\,dx$$
$$= \frac{2}{\pi}\sum_{k=1}^{n-1} \frac{1}{k} = \frac{2}{\pi}\left(1 + \frac{1}{2} + \frac{1}{3} + \cdots + \frac{1}{n-1}\right) \to \infty.$$

However, *the integral is convergent.* Since $\sin x/x$ is continuous on $[0, \pi/2]$, it suffices to prove convergence of the integral from $\pi/2$ to infinity. Integration by parts gives
$$\int_{\pi/2}^\infty \frac{\sin x}{x}\,dx = -\int_{\pi/2}^\infty \frac{1}{x}\,d[\cos x] = \left[\frac{\cos x}{x^2}\right]_{\pi/2}^\infty - \int_{\pi/2}^\infty \frac{\cos x}{x^2}\,dx = -\int_\pi^\infty \frac{\cos x}{x^2}\,dx.$$
But
$$\int_\pi^\infty \frac{\cos x}{x^2}\,dx$$
is absolutely convergent, since
$$\left|\frac{\cos x}{x^2}\right| \leq \frac{1}{x^2}.$$
(See Problem 2.) Thus,
$$\int_0^\infty \frac{\sin x}{x}\,dx$$
is a convergent improper integral. ∎

A congruent integral which is not absolutely convergent is said to be *conditionally convergent*.

5.10.1 Problems

1. Show that $\int_0^1 \frac{1}{x^p} dx$ converges absolutely iff $0 < p < 1$.

2. Show that $\int_1^\infty \frac{1}{x^p} dx$ converges absolutely iff $p > 1$.

3. Show that
$$\int_0^\infty e^{-ax} dx = 1/a.$$

4. Show that the *Fresnel integral*
$$\int_0^\infty \sin\left(x^2\right) dx$$
is convergent. (*Hint:* $\sin\left(x^2\right) = (1/2x)\left[2x \sin\left(x^2\right)\right]$.)

5. Which of the following integrals are absolutely convergent? Conditionally convergent? Divergent?

    (*Warning:* Many of these integrals cannot be computed explicitly.)

    (a) $\int_0^\infty e^{-x^2/2} dx$

    (b) $\int_0^\infty e^{-x} \sin x \, dx$

    (c) $\int_2^\infty \frac{1}{x (\log x)^2} dx$

    (d) $\int_0^{1/2} \frac{1}{x (\log x)} dx$

    (e) $\int_1^\infty \frac{\log x}{x\sqrt{x^2 - 1}} dx$

    (f) $\int_1^\infty \log x \sin\left(\frac{1}{x}\right) dx$

    (g) $\int_0^\infty \sin^2\left(\frac{1}{x}\right) dx$

    (h) $\int_0^\infty \frac{1}{\log x (1 + x^2)} dx$

    (i) $\int_0^\infty \sqrt{x} \cos x^2 \, dx$

*(j) $\displaystyle\int_0^\infty e^{-x} \log\left(\cos^2 x\right) dx$

*(k) $\displaystyle\int_2^\infty \frac{\sin x}{\log x} dx$

*(l) $\displaystyle\int_0^\infty \frac{x+1}{x^{3/2}} \sin x \, dx$

6. For which values of the parameters $p > 0$ and $q > 0$ are the following integrals absolutely convergent? Conditionally convergent?

(a) $\displaystyle\int_0^\infty \sin x^p \, dx$

(b) $\displaystyle\int_0^\infty x^q \sin x^p \, dx$

(c) $\displaystyle\int_0^\infty \frac{\sin x}{x^p} dx$

(d) $\displaystyle\int_0^1 |\log x|^p \, dx$

(e) $\displaystyle\int_0^1 \frac{1}{x^p |\log x|^q} dx$

(f) $\displaystyle\int_2^\infty \frac{1}{x^q (\log x)^p} dx$

5.11 Supplementary Problems

*1. Is the function
$$f(x) = \sin\left(\frac{1}{x}\right)$$
with $f(0) = 0$, integrable on $[-1, 1]$?

2. Let $f(x)$ be continuous on $[0, 1]$. Evaluate
$$\lim_{n \to \infty} \prod_{k=1}^n e^{(1/n) f(k/n)}.$$

(*Answer:* $\exp \int_0^1 f(x) \, dx$.)

3. Find
$$\lim_{n \to \infty} e^{1/n^2} e^{2/n^2} e^{3/n^2} \cdots e^{n/n^2}.$$

4. Let $f(x)$ be continuous and nonvanishing on $[0,1]$. Find

$$\lim_{n\to\infty} \frac{1}{n} \log \left\{ f\left(\frac{1}{n}\right) f\left(\frac{2}{n}\right) f\left(\frac{3}{n}\right) \ldots f\left(\frac{n}{n}\right) \right\}$$
$$= \lim_{n\to\infty} \frac{1}{n} \log \left\{ \prod_{k=1}^{n} f\left(\frac{k}{n}\right) \right\}.$$

(Answer: $\int_0^1 \log f(x)\, dx$.)

5. Find

$$\lim \frac{1}{n} \log \left(\frac{(2n)!}{n! n^n} \right).$$

6. (*Arithmetic and Geometric Means*) The *arithmetic mean* is the usual "average" of a finite set of numbers $a_1, a_2, \ldots a_n$. It is defined to be

$$A_n = \frac{a_1 + a_2 + \ldots + a_n}{n}.$$

The *geometric mean* is

$$G_n = (a_1, a_2, \ldots a_n)^{1/n}.$$

Let $f(x)$ be continuous on $[0,1]$, and partition $[0,1]$ into equal intervals. Let

$$a_k = f(\frac{k}{n})$$

be the value of $f(x)$ at the *k*th partition point. Find

$$A = \lim_{n\to\infty} A_n \quad \text{and} \quad G = \lim_{n\to\infty} G_n.$$

7. Define a new type of "derivative" by

$$D^* f(x) = \lim_{h\to 0} \frac{f^2(x+h) - f^2(x)}{h}.$$

(a) Express $D^* f(x)$ in terms of $f(x)$ and $f'(x)$.

(b) Find a version of the Fundamental Theorem for evaluating

$$\int_a^b D^* f(x)\, dx.$$

8. (*The Beta function*) Define the *Beta function* (or the "*Eulerian integral of the first kind*") to be
$$B(a,b) = \int_0^1 (1-x)^{a-1} x^{b-1} dx.$$

(a) Show that
$$B(a,b) = B(b,a).$$

(b) Show that
$$B(a,b) = 2 \int_0^{\frac{\pi}{2}} (\sin t)^{2a-1} (\cos t)^{2b-1} dt.$$

(c) Show that
$$= \int_{-1}^1 (1-x)^{a-1}(1+x)^{b-1} dx = 2^{a+b-1} B(a,b).$$

(*Hint:* Let $x = 2s - 1$.)

(d) Express
$$J_n = \int_{-1}^1 \left(1-x^2\right)^n dx$$
in terms of $B(a,b)$.

(e) Show that
$$B(a,1) = \frac{1}{a}.$$

(f) Integrate by parts to show that
$$B(a,b) = \left(\frac{b-1}{a}\right) B(a+1, b-1).$$

(g) Prove that if n and m are positive integers, then
$$B(a,b) = \frac{(b-1)(b-2)\cdots 2 \cdot 1}{a \cdots (a+b-2) \ldots} B(a+b-1, 1) = \frac{(a-1)!\,(b-1)!}{(a+b-1)!}.$$

9. (*Beta function*) Show that
$$B(a,b) = \int_0^\infty \frac{s^{a-1}}{(1+s)^{a+b}} ds.$$

10. Let
$$F(x; a, b) = \int_0^x t^a (1+t)^b \, dt$$
where $a > 0, b > 0$.

(a) Show that
$$(a+1) F(x; a, b) + bF(x; a+1, b-1) = x^{a+1} (1+x)^b.$$

(b) Find $F(x; 10, 2)$.

11. (*Elliptic integrals*) If $|m| \le 1$, show that
$$\int_0^1 \frac{\sqrt{1-mt^2}}{\sqrt{1-t^2}} \, dt = \int_0^{\pi/2} \sqrt{1 - m \sin^2 \theta} \, d\theta$$

and
$$\int_0^1 \frac{1}{\sqrt{(1-t^2)(1-mt^2)}} \, dt = \int_0^{\pi/2} \frac{1}{\sqrt{1-m \sin^2 \theta}} \, d\theta.$$

12. (*Wallis's product*) Let
$$I_n = \int_0^{\pi/2} \sin^n(x) \, dx.$$

(a) Show that
$$I_n = \frac{n-1}{n} I_{n-2}$$

and
$$I_0 = \frac{\pi}{2} \quad \text{and} \quad I_1 = 1.$$

Deduce from this that
$$I_{2m} = \int_0^{\pi/2} \sin^{2m}(x) \, dx = \frac{2m-1}{2m} \frac{2m-3}{2m-2} \cdots \frac{1}{2} \frac{\pi}{2}$$
$$I_{2m+1} = \int_0^{\pi/2} \sin^{2m+1}(x) \, dx = \frac{2m}{2m+1} \frac{2m-2}{2m-1} \cdots \frac{2}{3}.$$

(b) Show that
$$0 \leq I_{n+1} < I_n$$
so that
$$1 \leq \frac{I_{2m}}{I_{2m+1}} \leq \frac{I_{2m-1}}{I_{2m+1}} = 1 + \frac{1}{2m}.$$
Prove that therefore
$$\lim \frac{I_{2m}}{I_{2m+1}} = 1.$$

(c) From Part (a), show that
$$\frac{\pi}{2} = \frac{2}{1}\frac{2}{3}\frac{4}{3}\frac{4}{5}\frac{6}{5}\frac{6}{7} \cdots \frac{2m}{2m-1}\frac{2m}{2m+1}\frac{I_{2m}}{I_{2m+1}}.$$
Conclude that
$$\frac{\pi}{2} = \lim_{m \to \infty} \left(\frac{2}{1}\frac{2}{3}\frac{4}{3}\frac{4}{5}\frac{6}{5}\frac{6}{7} \cdots \frac{2m}{2m-1}\frac{2m}{2m+1} \right)$$
$$= \frac{2}{1}\frac{2}{3}\frac{4}{3}\frac{4}{5}\frac{6}{5}\frac{6}{7} \cdots .$$
This result is called *Wallis's product* after John Wallis (1616–1703).

13. (*Euler–Mascheroni constant*) Let
$$s_n = \sum_{k=1}^{n} \frac{1}{k}.$$
By considering the area under the graph of $1/x$, as in the proof of the Integral test, show that
$$\sum_{k=2}^{n+1} \frac{1}{k} < \int_{1}^{n} \frac{1}{x} dx < \sum_{k=1}^{n} \frac{1}{k}$$
or
$$s_{n+1} - 1 < \log n < s_n.$$
Let $\gamma_n = s_n - \log n > 0$. Show that
$$\gamma_n - \gamma_{n+1} > 0$$
and hence, that
$$\gamma = \lim \gamma_n$$
exists. The number γ is known as the *Euler–Mascheroni constant*; its value is approximately 0.5772. It is not known whether γ is rational.

14. (*The Gamma function*) Define the Gamma function by
$$\Gamma(p) = \int_0^\infty x^{p-1} e^{-x} dx.$$

(a) For what values of p is the integral convergent?
(b) Show that $\Gamma(1) = 1$.
(c) Show that $\Gamma(p+1) = p\Gamma(p)$.
(d) If n is a positive integer, show that $\Gamma(n+1) = n!$.

15. (*Frulanni's integral*) Show that
$$\int_0^\infty \frac{e^{-ax} - e^{-bx}}{x} dx = \log\left(\frac{b}{a}\right).$$

16. (*Frulanni's integral*) Let $f(x)$ be continuous on $0 < x < \infty$. Show that if $f(0+) = \lim_{x \to 0+} f(x)$ and $f(\infty) = \lim_{x \to \infty} f(x)$ exist, then
$$\int_0^\infty \frac{f(ax) - f(bx)}{x} dx = [f(0+) - f(\infty)] \log\left(\frac{b}{a}\right).$$

Chapter 6

Infinite Series

6.1 Convergence

In ordinary English, the terms *'sequence'* and *'series'* are practically synonymous. In Mathematics, however, they are *entirely different*. A sequence a_n is an *infinite list* of numbers (or other objects). A *series*

$$\sum_{n=1}^{\infty} a_n = a_1 + a_2 + \cdots + a_n + \cdots$$

is an *infinite sum*, the sum of all the numbers of the sequence a_n.

Now, the sum of an infinite number of terms has no immediately obvious meaning. How can we make sense of an infinite sum? We must define it as a limit. The idea is simply to *add up the first n terms, and take the limit as n tends to infinity*.

To be precise, we define the *nth partial sum* s_n of the series to be the *sum of the first n terms*:

$$s_n = a_1 + a_2 + \cdots + a_n = \sum_{k=1}^{n} a_k.$$

The partial sums of a series form a sequence

$$s_1, s_2, \ldots, s_n, \ldots.$$

The limit of this sequence is the sum of the series.

Definition 1 The series $\sum_{n=1}^{\infty} a_n$ is *convergent* iff the limit

$$s = \lim s_n$$

of the sequence of partial sums exists. The number s is called the *sum of the series*, and we write

$$s = \sum_{n=1}^{\infty} a_n = a_1 + a_2 + \cdots + a_n + \cdots.$$

If a series is not convergent, it is said to be *divergent*.

In the special case where

$$\lim s_n = \pm \infty,$$

the series is still said to be divergent, but we will write

$$\sum_{n=1}^{\infty} a_n = \pm \infty.$$

Example 1 As a simple explicit example, consider the series

$$\sum_{n=1}^{\infty} \frac{1}{n(n+1)} = \frac{1}{1 \cdot 2} + \frac{1}{2 \cdot 3} + \frac{1}{3 \cdot 4} + \cdots.$$

The partial sums are

$$s_1 = \frac{1}{2}$$
$$s_2 = \frac{1}{2} + \frac{1}{6} = \frac{2}{3}$$
$$s_2 = \frac{1}{2} + \frac{1}{6} + \frac{1}{12} = \frac{3}{4}$$
$$\vdots$$
$$s_n = \frac{n}{n+1}.$$

Hence,

$$\sum_{n=1}^{\infty} \frac{1}{n(n+1)} = \lim s_n = \lim \frac{n}{n+1} = 1. \quad \blacksquare$$

Theorem 1 (Linearity) *If $\sum_{n=1}^{\infty} a_n$ and $\sum_{n=1}^{\infty} b_n$ are convergent, then so are $\sum_{n=1}^{\infty} (a_n + b_n)$ and $\sum_{n=1}^{\infty} c\, a_n$, and we have*

$$\sum_{n=1}^{\infty} (a_n + b_n) = \sum_{n=1}^{\infty} a_n + \sum_{n=1}^{\infty} b_n$$

and
$$\sum_{n=1}^{\infty} c\, a_n = c \sum_{n=1}^{\infty} a_n.$$

Proof This follows immediately from the Basic Limit Theorems and is left as an exercise. □

Evaluation of Two Simple Series

One of the reasons that beginning students frequently dislike infinite series is that *it is usually difficult, or impossible, to find an explicit formula for the sum of a series*, much more so than is the case with derivatives or integrals.

However, one important case where summation is possible is the *Geometric Series*
$$\sum_{n=1}^{\infty} x^n = 1 + x + x^2 + x^3 + \cdots + x^n + \cdots.$$

Theorem 2 (Geometric Series) *The geometric series converges to*
$$\sum_{n=1}^{\infty} x^n = \frac{1}{1-x}$$
if $|x| < 1$, and diverges if $|x| \geq 1$.

Proof Note that
$$(1-x)\left(1 + x + x^2 + \cdots + x^n\right)$$
$$= 1 + x + x^2 + \ldots + x^n - x - x^2 - \cdots - x^{n+1}$$
$$= 1 - x^{n+1}.$$

For $x \neq 1$, the partial sum is therefore
$$s_n = 1 + x + x^2 + \cdots + x^n = \frac{1 - x^{n+1}}{1 - x}.$$

If $|x| < 1$, then $\lim x^{n+1} = 0$, and we have
$$\lim s_n = \lim \frac{1 - x^{n+1}}{1 - x} = \frac{1}{1-x}.$$

If $|x| \geq 1$, $\lim x^{n+1}$ does not exist, so the series diverges. □

Example 2 For example,
$$\sum_{n=1}^{\infty} \left(\frac{1}{3}\right)^n = 1 + \frac{1}{3} + \frac{1}{9} + \cdots = \frac{1}{1-\frac{1}{3}} = \frac{3}{2}.$$

Another simple case is that of a *telescoping sum*. ∎

Theorem 3 (Telescoping Sum) *If* $\lim a_n = a$, *then*
$$\sum_{n=1}^{\infty} (a_{n+1} - a_n) = a - a_1.$$

Proof We have for the partial sum
$$s_n = \sum_{n=1}^{n-1} (a_{k+1} - a_k)$$
$$= (a_2 - a_1) + (a_3 - a_2) + (a_4 - a_3) + \cdots + (a_n - a_{n-1})$$
$$= (a_n - a_1)$$

so that
$$\sum_{n=1}^{\infty} (a_{n+1} - a_n) = \lim s_n = \lim (a_n - a_1) = a - a_1. \qquad \square$$

Example 3 We have
$$\sum_{n=1}^{\infty} \frac{1}{n(n+1)} = \sum_{n=1}^{\infty} \left(\frac{1}{n} - \frac{1}{n+1}\right) = \lim \left(1 - \frac{1}{n+1}\right) = 1. \qquad \blacksquare$$

Convergence Tests

The number $\sum_{n=1}^{\infty} a_n$ is defined as a limit, and, as with all limits, there is a question of whether the limit exists. This is the question of *convergence of the series*. It amounts to asking whether in writing down the expression $\sum_{n=1}^{\infty} a_n$, we have written down a number or not. Thus, when confronting a series, it is a matter of first importance to determine whether it converges.

We shall therefore develop in subsequent sections a number of results that help to answer this question. We begin here with a simple necessary condition.

Theorem 4 (Necessary Condition for Convergence) *If $\sum_{n=1}^{\infty} a_n$ is convergent, then*

$$\lim a_n = 0.$$

Proof We have

$$a_n = s_n - s_{n-1} \to s - s = 0. \qquad \square$$

Example 4

(a) The series

$$\sum_{n=1}^{\infty} (-1)^{n+1} = 1 - 1 + 1 - 1 + \cdots$$

is divergent, because the *nth* term $(-1)^{n+1}$ does not tend to zero.

(b) The series

$$\sum_{n=1}^{\infty} n = 1 + 2 + 3 + \cdots + n + \cdots$$

is also divergent. Its *nth* term, n, does not tend to zero. In this case,

$$s_n = 1 + 2 + 3 + \ldots + n \geq 1 + 1 + \cdots + 1 = n \to \infty$$

so that

$$\sum_{n=1}^{\infty} n = \infty.$$

Explicitly, the partial sum is $s_n = n(n+1)/2$.

(c) Finally, the *nth* term of the series

$$\sum_{n=1}^{\infty} \frac{n}{n+1} = \frac{1}{2} + \frac{2}{3} + \frac{3}{4} + \frac{4}{5} + \cdots + \frac{n}{n+1} + \cdots$$

is $a_n = n/(n+1)$. The sequence a_n converges to the limit

$$\lim \frac{n}{n+1} = 1$$

but this is not zero, so again the series is divergent. ∎

The converse of Theorem 4 is false. It is definitely not true that a series necessarily converges if its nth term a_n tends to zero. It is quite possible for a series to be divergent even though its nth term tends to zero.

Example 5 The classic example is the *Harmonic series*

$$\sum_{n=1}^{\infty} \frac{1}{n} = 1 + \frac{1}{2} + \frac{1}{3} + \frac{1}{4} + \frac{1}{5} + \cdots + \frac{1}{n} + \cdots.$$

Here, clearly,

$$\lim \frac{1}{n} = 0,$$

but the series is none the less divergent, and in fact,

$$\sum_{n=1}^{\infty} \frac{1}{n} = \infty.$$

This is shown most easily by the Integral Test, as in the next section. However, the classic direct proof is as follows. Let

$$s_n = \sum_{k=1}^{n} \frac{1}{k} = 1 + \frac{1}{2} + \frac{1}{3} + \frac{1}{4} + \cdots + \frac{1}{n}.$$

Then s_n is an increasing sequence. We have

$$s_{2^n} = \sum_{n=1}^{2^n} \frac{1}{n} = 1 + \frac{1}{2} + \frac{1}{3} + \frac{1}{4} + \frac{1}{5} + \frac{1}{6} + \frac{1}{7} + \frac{1}{8} + \cdots + \frac{1}{2^n}$$

$$= 1 + \frac{1}{2} + \left(\frac{1}{3} + \frac{1}{4}\right) + \left(\frac{1}{5} + \frac{1}{6} + \frac{1}{7} + \frac{1}{8}\right) + \cdots + \left(\frac{1}{1+2^{n-1}} + \cdots + \frac{1}{2^n}\right)$$

$$> 1 + \frac{1}{2} + \left(\frac{1}{4} + \frac{1}{4}\right) + \left(\frac{1}{8} + \frac{1}{8} + \frac{1}{8} + \frac{1}{8}\right) + \cdots + \left(\frac{1}{2^n} + \cdots + \frac{1}{2^n}\right)$$

$$= 1 + \frac{1}{2} + \frac{1}{2} + \frac{1}{2} + \cdots + \frac{1}{2} = 1 + \frac{n}{2} \to \infty.$$

The sequence s_n of partial sums is unbounded, so the series diverges. ∎

6.1.1 Problems

1. In Example 1, prove by induction that

$$s_n = \frac{n}{n+1}.$$

6.2 Series of Positive Terms

We next consider convergence for the important case in which *all terms of the series are positive*. We say that

$$\sum_{n=1}^{\infty} a_n$$

is a *series of positive terms* iff $a_n \geq 0$ for all n.

The following fact is fundamental.

Theorem 5 *A series of positive terms is convergent iff its partial sums are bounded.*

Proof Since we have

$$s_{n+1} = (a_1 + a_2 + \ldots + a_n) + a_{n+1} \geq a_1 + a_2 + \ldots + a_n = s_n$$

the sequence of partial sums is increasing. According to the *Monotone Sequence Theorem*, the increasing sequence s_n converges iff it is bounded. □

By Theorem 5 the increasing sequence s_n is either bounded and converges, or is unbounded and tends to infinity. Hence, formally, $\sum_{n=1}^{\infty} a_n$ has a meaning as either a finite number or ∞, according to whether it converges or not. Thus, the statement

$$\sum_{n=1}^{\infty} a_n < \infty$$

for a sequence of positive terms, simply means that $\sum_{n=1}^{\infty} a_n$ converges.

Theorem 5 is the basis of three basic tests for convergence: the *Integral Test*, the *Comparison Test*, and the *Limit Comparison Test*.

Theorem 6 (Integral Test) *Let $f(x)$ be a positive decreasing function on $[1, \infty)$. If $a_n = f(n)$ for all integers $n \geq 1$, then*

$$\sum_{n=1}^{\infty} a_n < \infty$$

iff

$$\int_1^{\infty} f(x) dx < \infty. \tag{6.1}$$

Proof For $n \leq x \leq n+1$, we have
$$a_n = f(n) \geq f(x) \geq f(n+1) \geq a_{n+1}.$$
Integrating this over the interval $[n, n+1]$ gives
$$a_n = \int_n^{n+1} a_n \, dx \geq \int_n^{n+1} f(x) \, dx \geq \int_n^{n+1} a_{n+1} \, dx = a_{n+1}.$$
Summing from $n = 1$ to m gives
$$s_m = \sum_{n=1}^m a_n \geq \sum_{n=1}^m \int_n^{n+1} f(x)dx = \int_1^{m+1} f(x)dx \geq \sum_{n=1}^m a_{n+1} = s_m - a_1$$
or
$$s_m \geq \int_1^{m+1} f(x)dx \geq s_m - a_1.$$

Thus if (6.1) holds, then the sequence s_m is bounded by
$$\int_1^\infty f(x)dx.$$
Since s_n is increasing it must converge.

Conversely, if s_m is bounded by M, then
$$\int_1^b f(x)dx \leq \int_1^{[b]+1} f(x)dx \leq s_{[b]} \leq M$$
where $[b]$ is the greatest integer in b, so that
$$\int_1^\infty f(x)dx$$
also converges. $\square$

Corollary 1 *The series*
$$\sum_{n=1}^\infty \frac{1}{n^p} < \infty$$
iff $p > 1$.

Proof The result follows from Theorem 1, since for $p > 1$
$$\int_1^b \frac{1}{x^p} dx = \lim_{b \to \infty} \int_1^b \frac{1}{x^p} dx = \frac{1}{p-1} \lim_{b \to \infty} \left(1 - \frac{1}{b^{p-1}}\right) = \frac{1}{p-1}.$$

Hence, the series is convergent if $p > 1$.
Conversely, if $p \leq 1$,

$$\int_1^b \frac{1}{x^p} dx = \infty$$

and the series diverges. □

Theorem 7 (Comparison Test) *Suppose that $0 \leq a_n \leq b_n$ for all $n \geq 1$. Then*

(a) *If $\sum_{n=1}^\infty b_n$ is convergent, then $\sum_{n=1}^\infty a_n$ is convergent.*

(b) *If $\sum_{n=1}^\infty a_n$ is divergent, then $\sum_{n=1}^\infty b_n$ is divergent.*

Proof For (a), we have

$$\sum_{n=1}^m a_n \leq \sum_{n=1}^m b_n \leq \sum_{n=1}^\infty b_n$$

so that the partial sums are bounded.

For (b), if $\sum_{n=1}^\infty b_n$ were convergent, then $\sum_{n=1}^\infty a_n$ would also be convergent by (a), which by hypothesis it is not. □

Example 1 Consider

$$\sum_{n=1}^\infty \frac{n}{n^3 + n + 1}.$$

We have

$$\frac{n}{n^3 + n + 1} \leq \frac{n}{n^3} = \frac{1}{n^2}.$$

Since

$$\sum_{n=1}^\infty \frac{1}{n^2} < \infty$$

we must therefore also have that

$$\sum_{n=1}^\infty \frac{n}{n^3 + n + 1} < \infty.$$ ■

We say that a_n is *asymptotic to* b_n iff
$$\lim \frac{a_n}{b_n} = 1.$$
In this case, we write
$$a_n \sim b_n.$$

Theorem 8 (Limit Comparison Test) *If a_n and b_n are positive and*
$$a_n \sim b_n,$$
then $\sum_{n=1}^{\infty} a_n$ converges iff $\sum_{n=1}^{\infty} b_n$ converges.

Proof Suppose that $\sum_{n=1}^{\infty} b_n$ converges. Since a_n/b_n converges, it is bounded. Hence, for some M,
$$\frac{a_n}{b_n} \leq M$$
or
$$a_n \leq M \, b_n.$$
Hence, $\sum_{n=1}^{\infty} a_n$ converges by the comparison test.
The converse follows by interchanging a_n and b_n. □

Example 2 Consider again
$$\sum_{n=1}^{\infty} \frac{n}{n^3 + n + 1}.$$
We have
$$\frac{n}{n^3 + n + 1} \sim \frac{1}{n^2}$$
so again the series converges. ∎

Important Remarks

We now ask a fundamental question. What exactly is the difference between the two series
$$\sum_{n=1}^{\infty} \frac{1}{n^2}$$

and
$$\sum_{n=1}^{\infty} \frac{1}{n} ?$$

Why does one converge and the other not, even though the nth term tends to zero in both cases?

The answer is that *what is at issue in the convergence of a series of positive terms is not whether the nth term goes to zero* (although that is certainly necessary), *but how fast it goes to zero.*

For example, $1/n^2$ goes to zero sufficiently faster than $1/n$ that

$$\sum_{n=1}^{\infty} \frac{1}{n^2}$$

is convergent, while

$$\sum_{n=1}^{\infty} \frac{1}{n}$$

is not.

Thus, in the *Comparison Test*, the condition that $0 \leq a_n \leq b_n$ implies that a_n goes to zero at least as fast as b_n, and so $\sum_{n=1}^{\infty} a_n$ must converge whenever $\sum_{n=1}^{\infty} b_n$ does.

Similarly, two sequences that are asymptotic go to zero *at the same rate*, and so the *Limit Comparison Test* simply states that if the *nth* terms of two series go to zero at the same rate, then one series converges iff the other does.

Once it is realized that this is the issue, the mystification usually experienced by students when confronted by the standard battery of convergence tests rapidly disappears.

6.2.1 Problems

1. Prove that if $a_n > 0$ and $\sum_{n=1}^{\infty} a_n$ diverges, then so does
$$\sum_{n=1}^{\infty} \frac{a_n}{1+a_n}.$$

2. Prove that if $a_n > 0$ and $\sum_{n=1}^{\infty} a_n$ converges, then so does
$$\sum_{n=1}^{\infty} \frac{\sqrt{a_n}}{n}.$$

(*Hint:* Use $2xy \leq x^2 + y^2$.)

3. Which of the following series converge?

   (a) $\sum_{n=0}^{\infty} \dfrac{2^n}{3^n + n}$

   (b) $\sum_{n=1}^{\infty} \dfrac{n^2 + 1}{n^4 + 2}$

   (c) $\sum_{n=1}^{\infty} \dfrac{n^3 + 1}{n^4 - 2}$

   (d) $\sum_{n=1}^{\infty} \dfrac{1}{n \log n \, (\log n)^2}$

   (e) $\dfrac{1}{1 \cdot 2 \cdot 3} + \dfrac{1}{2 \cdot 3 \cdot 4} + \dfrac{1}{3 \cdot 4 \cdot 5} + \cdots$

4. For what values of $p > 0$ does the series

$$\sum_{n=1}^{\infty} \dfrac{1}{n (\log n)^p}$$

   converge?

5. Adapt the method of Example 5 of Section 6.1 to prove that the series

$$\sum_{n=1}^{\infty} \dfrac{1}{n^2}$$

   converges to a number ≤ 2.5. (The actual value is $\pi^2/6 = 1.645$.)

6.3 The Ratio and Root Tests

We shall prove two standard tests for convergence that will be important in Chapter 8 for Power Series.

Theorem 9 (Ratio Test) *Let $a_n > 0$.*

(a) *If*

$$\lim \dfrac{a_{n+1}}{a_n} < 1,$$

then $\sum_{n=1}^{\infty} a_n$ converges.

(b) *If*
$$\lim \frac{a_{n+1}}{a_n} > 1,$$
then $\sum_{n=1}^{\infty} a_n$ *diverges.*

Proof

(a) Choose θ such that
$$\lim \frac{a_{n+1}}{a_n} < \theta < 1.$$
Then, for some N, $n \geq N$ implies that
$$\frac{a_{n+1}}{a_n} < \theta.$$
Hence, for $n \geq N$,
$$a_n \leq \theta\, a_{n-1} \leq \theta^2\, a_{n-2} \leq \ldots \leq \theta^{n-N} a_N.$$
Thus a_n is dominated by the convergent geometric series
$$a_N \theta^{-N} \sum_{n=1}^{\infty} \theta^n$$
and is therefore convergent.

(b) Choose θ such that
$$\lim \frac{a_{n+1}}{a_n} > \theta > 1.$$
Then for some N, $n \geq N$ implies
$$\frac{a_{n+1}}{a_n} > \theta.$$
Hence, for $n \geq N$,
$$a_n \geq \theta\, a_{n-1} \geq \theta^2\, a_{n-2} \geq \ldots \geq \theta^{n-N} a_N.$$
Thus a_n is unbounded so the series must diverge. □

Remark If the limiting ratio is 1; that is, if
$$\lim \frac{a_{n+1}}{a_n} = 1,$$

the test fails. *No conclusion can then be drawn.* This is well illustrated by the series

$$\sum_{n=1}^{\infty} \frac{1}{n^p}$$

where the ratio tends to 1 for every p, although the series converges for $p > 1$, and diverges for $0 < p < 1$.

In fact, as the proof shows, the Ratio Test applies only to series in which the nth term goes to zero *exponentially*, which is quite rapid convergence. Thus the Ratio Test, although useful, is a *fairly insensitive test for convergence*.

Theorem 10 (Root Test) *Let $a_n > 0$.*

(a) *If*

$$\lim a_n^{1/n} < 1,$$

then $\sum_{n=1}^{\infty} a_n$ converges.

(b) *If*

$$\lim a_n^{1/n} > 1,$$

then $\sum_{n=1}^{\infty} a_n$ diverges.

Proof

(a) Choose θ such that

$$\lim a_n^{1/n} < \theta < 1.$$

Then for some N, $n \geq N$ implies

$$a_n^{1/n} < \theta$$

or

$$a_n < \theta^n$$

if $n \geq N$, so that a_n is again dominated by a convergent geometric series.

(b) Choose θ such that

$$\lim a_n^{1/n} > \theta > 1.$$

Then for some N, $n \geq N$ implies

$$a_n^{1/n} > \theta$$

or

$$a_n > \theta^n.$$

Thus, again, a_n tends to infinity exponentially, and the series diverges.

□

Like the Ratio Test, the Root Test gives convergence only for series whose terms go to zero exponentially.

Note also that by Problem 3 of Section 2.10, if a_{n+1}/a_n converges to a limit L, then $a_n^{1/n}$ also converges to L, but not conversely. Thus the Root Test is somewhat more general than the Ratio Test.

6.3.1 Problems

1. Which of the following series converge?

    (a) $\sum_{n=0}^{\infty} \dfrac{(n!)^2}{(2n)!}$

    (b) $\sum_{n=0}^{\infty} \dfrac{1}{n!}$

    (c) $\sum_{n=0}^{\infty} \dfrac{1}{n^4} \dfrac{3^n}{2^n}$

2. For what $p > 0$ does the series
$$\sum_{n=1}^{\infty} \frac{1}{n^{2p}} p^{np}$$
converge?

3. Show that the Root Test works for the series
$$\sum_{n=0}^{\infty} \frac{1}{2^n} e^{(-1)^n \sqrt{n}}$$
but the Ratio Test does not.

6.4 Absolute and Conditional Convergence

We next turn to series whose terms are not necessarily positive.

Absolute Convergence

Definition 2 The series $\sum_{n=1}^{\infty} a_n$ *converges absolutely* iff the series
$$\sum_{n=1}^{\infty} |a_n|$$
of absolute values converges.

Theorem 11 *If a series converges absolutely, then it converges.*

Proof 1 Let
$$\sigma_n = |a_1| + |a_2| + \ldots + |a_n|$$
be the partial sums of the series of absolute values. Since $\sum_{n=1}^{\infty} |a_n|$ is convergent, σ_n is a Cauchy sequence. Thus for $m > n$,
$$|s_m - s_n| = \left|\sum_{k=n+1}^{m} a_k\right| \leq \sum_{k=n+1}^{m} |a_k| = \sigma_m - \sigma_n \to 0.$$
Hence, s_n is a Cauchy sequence as well. □

For our second proof, we will need a definition.

Definition 3 (Positive and Negative Parts of a Series) Let $\sum_{n=1}^{\infty} a_n$ be any series, convergent or not. Define the nonnegative sequences P_n and N_n by
$$P_n = \begin{cases} a_n & \text{if } a_n \geq 0 \\ 0 & \text{if } a_n < 0 \end{cases}$$
and
$$N_n = \begin{cases} |a_n| & \text{if } a_n < 0 \\ 0 & \text{if } a_n \geq 0. \end{cases}$$

These are simply the sequences of absolute values of positive and negative terms of the series $\sum_{n=1}^{\infty} a_n$. For example, for the series
$$\sum_{n=1}^{\infty} \frac{(-1)^{n+1}}{n} = 1 - \frac{1}{2} + \frac{1}{3} - \frac{1}{4} + \cdots$$
the sequence P_n is
$$1, 0, \frac{1}{3}, 0, \frac{1}{5}, \ldots$$
while N_n is
$$0, \frac{1}{2}, 0, \frac{1}{4}, 0, \frac{1}{6}, \ldots.$$

Lemma 1

(a) *We have*
$$a_n = P_n - N_n$$
and
$$|a_n| = P_n + N_n.$$

(b) *We have*
$$P_n = \frac{1}{2}(|a_n| + a_n)$$
and
$$N_n = \frac{1}{2}(|a_n| - a_n).$$

Proof Part (a) is clear, and Part (b) follows from it by simple algebra. □

We are now ready for our second proof of Theorem 11.

Proof 2 If $\sum_{n=1}^{\infty} |a_n|$ is convergent, then so are $\sum_{n=1}^{\infty} P_n$ and $\sum_{n=1}^{\infty} N_n$ by the comparison test, since
$$P_n \leq |a_n|$$
and
$$N_n \leq |a_n|.$$
Hence,
$$\sum_{n=1}^{\infty} a_n = \sum_{n=1}^{\infty} P_n - \sum_{n=1}^{\infty} N_n$$
is also convergent. □

Conditional Convergence

Definition 4 *The series $\sum_{n=1}^{\infty} a_n$ converges conditionally iff it converges, but does not converge absolutely.*

Theorem 12 *If $\sum_{n=1}^{\infty} a_n$ converges conditionally, then the series*

$$\sum_{n=1}^{\infty} P_n$$

and

$$\sum_{n=1}^{\infty} N_n$$

are both divergent and

$$\lim P_n = \lim N_n = 0.$$

Proof By Lemma 1, each of these series is the sum of a convergent series and a divergent series, and is therefore divergent. Moreover, both tend to zero, since a_n tends to zero. □

Alternating Series

A series whose terms alternate in sign is called an alternating series. Thus an *alternating series* is one of the form

$$\sum_{n=1}^{\infty} (-1)^{n+1} a_n$$

where $a_n > 0$.

Theorem 13 (Leibnitz) *Let $a_n > a_{n+1} \geq 0$ be a decreasing sequence of positive terms such that*

$$\lim a_n = 0.$$

Then the alternating series

$$\sum_{n=1}^{\infty} (-1)^{n+1} a_n = a_1 - a_2 + a_3 - a_4 + \cdots$$

is convergent.

Proof We have

$$s_{2n+1} = s_{2n-1} - (a_{2n} - a_{2n+1}) < s_{2n-1}$$

and
$$s_{2n+2} = s_{2n} + (a_{2n+1} - a_{2n+2}) > s_{2n}.$$
Thus, the sequence of *even* terms is increasing, and the sequence of *odd* terms is decreasing. By the Monotone Sequence Theorem,
$$\lim s_{2n} = s$$
exists. But
$$\lim s_{2n+1} = \lim s_{2n} + a_{2n+1} = s + 0 = s.$$
Hence, both sequences converge to the same limit. It follows that $\lim s_n = s$. □

Example 1 This shows that the series
$$1 - \frac{1}{2} + \frac{1}{3} - \frac{1}{4} + \cdots = \sum_{n=1}^{\infty} (-1)^{n+1} \frac{1}{n}$$
is convergent, but not absolutely convergent. The value of this series will be shown in Chapter 9 to be $\log 2$. ∎

Example 2 In Leibnitz's Test, it is essential that the sequence a_n be *monotone*. For example, the alternating series
$$1 - \frac{1}{2^2} + \frac{1}{3} - \frac{1}{4^2} + \cdots = \sum_{n=1}^{\infty} (-1)^{n+1} a_n$$
where
$$a_n = \begin{cases} 1/n & n \text{ odd} \\ 1/n^2 & n \text{ even} \end{cases}$$
has $\lim a_n = 0$, but is divergent because its series of positive terms diverges, while its series of negative terms converges. ∎

6.4.1 Problems

1. Classify the following series as absolutely convergent, conditionally convergent, or divergent.

   (a) $\sum_{n=1}^{\infty} (-1)^{n+1} \frac{n}{\log n}$

(b) $\sum_{n=1}^{\infty} (-1)^{n+1} \dfrac{\log n}{n}$

(c) $\sum_{n=1}^{\infty} (-1)^{n+1} \dfrac{\log n}{n^2}$

(d) $\sum_{n=1}^{\infty} (-1)^{n+1} \dfrac{2^n}{\sqrt{n!}}$

(e) $\sum_{n=1}^{\infty} \dfrac{(-1)^{n+1}}{\sqrt{n}}$

(f) $1 - \dfrac{1}{3} + \dfrac{1}{3^2} - \dfrac{1}{3^3} + \dfrac{1}{3^4} - \cdots$

(g) $\dfrac{1}{\sqrt{1}} + \dfrac{1}{\sqrt{3}} - \dfrac{1}{\sqrt{2}} + \dfrac{1}{\sqrt{5}} + \dfrac{1}{\sqrt{7}} - \dfrac{1}{\sqrt{4}} + \dfrac{1}{\sqrt{9}} + \dfrac{1}{\sqrt{11}} - \dfrac{1}{\sqrt{6}} + \cdots$

(h) $1 + \dfrac{1}{3} - \dfrac{1}{2} + \dfrac{1}{5} + \dfrac{1}{7} - \dfrac{1}{4} + \dfrac{1}{9} + \dfrac{1}{11} - \dfrac{1}{6} + \cdots$

(*Hint for* (g): Bracket the terms three at a time.)

2. In Leibnitz's Test, since the sequence s_{2n} of even terms is strictly increasing and the sequence of s_{2n+1} of odd terms is strictly decreasing, the sum s lies between successive partial sums; that is,

$$s_{2n} < s < s_{2n+1}.$$

(a) Show that if s is approximated by the partial sum s_n, the error is less than the next term a_{n+1}.

(b) Use this to compute

$$\dfrac{1}{e} = \sum_{n=0}^{\infty} (-1)^n \dfrac{1}{n!}$$

correct to five decimal places. (*Answer:* 0.36788.)

6.5 *Rearrangement of Series

An important difference between absolutely and conditionally convergent series is their behavior when their terms are rearranged in a different order. *The rearrangement of a conditionally convergent series may converge to a different value than the original series.* A classic example is the conditionally convergent series

$$\log 2 = 1 - \dfrac{1}{2} + \dfrac{1}{3} - \dfrac{1}{4} + \cdots.$$

6.5 *Rearrangement of Series

We have
$$\frac{1}{2}\log 2 = \left(\frac{1}{2} - \frac{1}{4}\right) + \left(\frac{1}{6} - \frac{1}{8}\right) + \cdots$$
$$= \left(0 + \frac{1}{2} + 0 - \frac{1}{4}\right) + \left(0 + \frac{1}{6} + 0 - \frac{1}{8}\right) + \cdots.$$

Therefore
$$\frac{3}{2}\log 2 = \log 2 + \frac{1}{2}\log 2$$
$$= \left(1 - \frac{1}{2} + \frac{1}{3} - \frac{1}{4}\right) + \left(\frac{1}{5} - \frac{1}{6} + \frac{1}{7} - \frac{1}{8}\right) + \cdots$$
$$+ \left(0 + \frac{1}{2} + 0 - \frac{1}{4}\right) + \left(0 + \frac{1}{6} + 0 - \frac{1}{8}\right) + \cdots$$
$$= \left(1 + 0 + \frac{1}{3} - \frac{1}{2}\right) + \left(\frac{1}{5} + 0 + \frac{1}{7} - \frac{1}{4}\right) + \cdots$$
$$= \left(1 + \frac{1}{3} - \frac{1}{2}\right) + \left(\frac{1}{5} + \frac{1}{7} - \frac{1}{4}\right) + \cdots.$$

This is a rearrangement of the original series for $\log 2$.

A rearrangement of a conditionally convergent series may also diverge. For example, the series
$$\sum_{n=1}^{\infty} \frac{(-1)^{n+1}}{\sqrt{n}} = \frac{1}{\sqrt{1}} - \frac{1}{\sqrt{2}} + \frac{1}{\sqrt{3}} - \frac{1}{\sqrt{4}} + \cdots + \frac{(-1)^{n+1}}{\sqrt{n}} + \cdots$$
converges, but its rearrangement
$$\left(\frac{1}{\sqrt{1}} + \frac{1}{\sqrt{3}} - \frac{1}{\sqrt{2}}\right) + \left(\frac{1}{\sqrt{5}} + \frac{1}{\sqrt{7}} - \frac{1}{\sqrt{4}}\right) + \left(\frac{1}{\sqrt{9}} + \frac{1}{\sqrt{11}} - \frac{1}{\sqrt{6}}\right) + \cdots$$
diverges. (See Problem 1.)

This cannot happen if the series is absolutely convergent.

Theorem 14 *If $\sum_{n=0}^{\infty} a_n$ is absolutely convergent to s, then every rearrangement of $\sum_{n=0}^{\infty} a_n$ converges to s.*

Proof Let $\sum_{n=0}^{\infty} b_n$ be a rearrangement of $\sum_{n=0}^{\infty} a_n$. Let $\epsilon > 0$, and choose N such that
$$\sum_{n=N}^{\infty} |a_n| < \frac{\epsilon}{3}.$$

Choose an $M \geq N$ such that all the terms $a_1, \ldots, a_N$ occur in the list $b_1, \ldots, b_M$. If $n \geq M$, then in the sum

$$\sum_{k=0}^{n} b_k - \sum_{k=0}^{n} a_k$$

all the terms $a_1, \ldots, a_N$ cancel out, and we are left with two sums of terms a_k with $k > N$. Thus

$$\left| \sum_{k=0}^{n} b_k - \sum_{k=0}^{n} a_k \right| \leq 2 \sum_{k=N+1}^{\infty} a_k < \frac{2}{3}\epsilon$$

and hence

$$\left| \sum_{k=0}^{n} b_k - s \right| \leq \left| \sum_{k=0}^{n} b_k - \sum_{k=0}^{n} a_k \right| + \left| \sum_{k=n+1}^{\infty} a_k \right| < \frac{2}{3}\epsilon + \frac{1}{3}\epsilon = \epsilon. \qquad \square$$

On the other hand, a conditionally convergent series can be rearranged to give any value whatever, including $\pm\infty$.

Theorem 15 *If $\sum_{n=0}^{\infty} a_n$ is conditionally convergent and α is any number (including $\pm\infty$), then there is a rearrangement of $\sum_{n=0}^{\infty} a_n$ that converges to α.*

Proof We shall prove the result for α finite. We construct a rearrangement b_n of a_n as follows. Starting with the positive term $b_0 = P_0$, we let $b_k = P_k$ until the first time that

$$b_0 + b_1 + \cdots + b_{n_1} = P_0 + P_1 + \cdots + P_{n_1} > \alpha.$$

Then we take $b_k = N_{k-n_1}$, putting in enough negative terms until the first time that

$$b_0 + b_1 + \cdots + b_{n_2} = P_0 + P_1 + \cdots - N_{n_2} < \alpha.$$

We continue in this manner. We note that since both positive and negative series diverge to ∞, it is always possible to add in enough terms to surpass any number whatever. On the other hand, because the series is convergent, $a_n \to 0$, and hence also $P_n \to 0$ and $N_n \to 0$. Since we always change from positive to negative terms the first time the partial sum passes a, the partial sum will differ from a by no more than the last term of the series, which tends to zero. $\qquad \square$

6.5.1 Problems

1. Show that the series
$$\sum_{n=1}^{\infty} \frac{(-1)^{n+1}}{\sqrt{n}} = \frac{1}{\sqrt{1}} - \frac{1}{\sqrt{2}} + \frac{1}{\sqrt{3}} - \frac{1}{\sqrt{4}} + \cdots + \frac{(-1)^{n+1}}{\sqrt{n}} + \cdots$$
converges, but that its rearrangement
$$\left(\frac{1}{\sqrt{1}} + \frac{1}{\sqrt{3}} - \frac{1}{\sqrt{2}}\right) + \left(\frac{1}{\sqrt{5}} + \frac{1}{\sqrt{7}} - \frac{1}{\sqrt{4}}\right) + \left(\frac{1}{\sqrt{9}} + \frac{1}{\sqrt{11}} - \frac{1}{\sqrt{6}}\right) + \cdots$$
diverges.

2. Prove Theorem 15 for $\alpha = \infty$.

Chapter 7

Uniform Convergence

7.1 Limits of Sequences of Functions

Consider a sequence of functions $f_n(x)$, which converges to a limit function $f(x)$:

$$\lim f_n(x) = f(x). \tag{7.1}$$

Suppose that we are interested in deducing various properties of the limit function from similar properties of the functions $f_n(x)$ of the sequence. For example:

1. If all the $f_n(x)$ are continuous, is $f(x)$ continuous?

2. If all the $f_n(x)$ are differentiable, is $f(x)$ differentiable, and, if so, is

$$f'(x) = \lim f'_n(x)? \tag{7.2}$$

3. Is it true that

$$\lim \int_a^b f_n(x)dx = \int_a^b f(x)dx?$$

The answer to all these questions is easily given. It is a resounding "*No!*" If all that is meant by (7.1) is that $f_n(x) \to f(x)$ *for each fixed x*, then none of these things need be true.

Here are some examples.

Example 1 Consider the sequence of functions $f_n(x) = x^n$ on $0 \le x \le 1$. These functions are all continuous, even differentiable, but the limit function

$$f(x) = \lim f_n(x) = \begin{cases} 0 & 0 \le x < 1 \\ 1 & x = 1 \end{cases}$$

is discontinuous. ∎

Example 2 The sequence of functions

$$f_n(x) = \frac{\sin(n^2 x)}{n}$$

converges to zero, but the limit of its derivative

$$f_n'(x) = n \cos(n^2 x)$$

does not exist. ∎

Example 3

(a) For the sequence of functions

$$s_n(x) = \begin{cases} n & 0 < x < \frac{1}{n} \\ 1 & \text{otherwise} \end{cases}$$

on the interval $0 \le x < 1$,

$$s(x) = \lim s_n(x) = 0$$

for all x. But

$$\lim \int_0^1 s_n(x) dx = 1 \ne \int_0^1 s(x) dx = 0.$$

(b) The sequence of functions

$$f_n(x) = nxe^{-nx}$$

converges to zero on $x \ge 0$, but

$$\int_0^1 f_n(x) dx = 1$$

for all n. ∎

The sort of convergence considered above is called *"pointwise convergence."* We say that *the sequence $f_n(x)$ converges to $f(x)$ pointwise on D* iff for each point $x \in D$,

$$\lim f_n(x) = f(x).$$

Pointwise convergence is a very weak sort of convergence. We need to find some *extra properties* that a sequence of functions must have in order that these or other properties will be preserved. This subject is enormously complicated. There is an immense number of different conditions on a sequence that ensure the preservation of various properties in the limit. We shall, however, study only one of these, the simplest and most useful, which is known as *Uniform Convergence.*

7.2 Uniform Convergence

We consider a sequence of functions $f_n(x)$, which converge pointwise to a limit function $f(x)$.

Definition 1 *The sequence $f_n(x)$ converges to $f(x)$ uniformly on a set D iff for every $\epsilon > 0$, there is a positive integer N, such that for all $x \in D$,*

$$|f_n(x) - f(x)| < \epsilon$$

whenever $n \geq N$.

The crucial point is that *the same N works for all x at the same time*. This may seem to be a very technical point, and so it is. This is a technical subject, but a very important one.

What the definition says is that if n is large,

$$f(x) - \epsilon < f_n(x) < f(x) + \epsilon.$$

This means that the graph of $f_n(x)$ lies in a strip of width 2ϵ about the graph of $f(x)$. That is to say, the graph of the function $f_n(x)$ is close to that of $f(x)$ everywhere on the set D.

A crucial point is that uniform convergence is defined *on some set D*. Thus *a given sequence may converge uniformly on one set, but not on another.*

A good way to show that $f_n(x) \to f(x)$ uniformly is to estimate the difference by something independent of x that goes to zero, as in the next Lemma.

Lemma 1 *If $|f_n(x) \to f(x)| \leq c_n$ for all $x \in D$, and $\lim c_n = 0$, then $f_n(x) \to f(x)$ uniformly on D.*

Proof Let $\epsilon > 0$, and choose N such that $c_n < \epsilon$ for $n \geq N$. Then $|f_n(x) \to f(x)| < \epsilon$ for $n \geq N$. $\square$

In order to grasp what is going on, we need to consider some examples. The reader is advised to sketch the graphs of these functions to see more clearly what is happening.

Example 1 The sequence of functions $f_n(x) = x^n$ converges on the interval $[0,1]$ to the function

$$f(x) = \begin{cases} 0 & \text{if } 0 \leq x < 1 \\ 1 & \text{if } x = 1 \end{cases}$$

but the convergence is not uniform on the interval $[0,1]$. However, it is uniform on any smaller interval of the form $[0,a]$ with $0 < a < 1$, since there we have

$$0 \leq x^n \leq a^n \to 0. \qquad \blacksquare$$

Example 2 The sequence of functions

$$f_n(x) = \frac{\sin n^2 x}{n}$$

converges uniformly to 0 on $[0, \pi]$, since

$$|f_n(x)| \leq \frac{1}{n} \to 0.$$

However, the sequence of its derivatives

$$f_n'(x) = n \cos n^2 x$$

does not converge at any point. $\blacksquare$

Example 3 The sequence of functions

$$f_n(x) = nxe^{-nx}$$

converges to zero on $x \geq 0$, but not uniformly, since

$$f_n(1/n) = e^{-1}.$$

However, $f_n(x)$ does converge uniformly on any interval $0 < a \leq x < \infty$. For $f_n(x)$ attains its maximum value at $x = 1/n$, and $f_n'(x) < 0$ for $x > 1/n$. Hence, if $n > 1/a$, we have

$$0 \leq f_n(x) \leq f_n(a) = nae^{-na} \to 0$$

for $a \leq x < \infty$. $\blacksquare$

7.2.1 Problems

1. Prove that
$$f_n(x) = \frac{x^2}{x^2 + (nx-1)^2}$$
converges to zero on $0 \leq x \leq 1$, but not uniformly. (*Hint:* What is $f_n(1/n)$?)

2. Prove that if $f_n(x)$ and $g_n(x)$ converge uniformly on a set S, then
$$f_n(x) + g_n(x)$$
converges uniformly on S.

3. Prove that if $f_n(x)$ and $g_n(x)$ converge uniformly on a set S, and $f_n(x)$ and $g_n(x)$ are *bounded*, then $f_n(x)g_n(x)$ converges uniformly on S.

4. Give an example of sequences $f_n(x)$ and $g_n(x)$ converging uniformly on a set S, for which $f_n(x)g_n(x)$ does not converge uniformly on S.

7.3 Continuity

The uniform limit of continuous functions is continuous.

Theorem 1 *Let $f_n(x)$ converge to $f(x)$ uniformly on the interval $[a,b]$. If the functions $f_n(x)$ are continuous on $[a,b]$, then $f(x)$ is continuous on $[a,b]$.*

Proof Fix a point c in $[a,b]$. We shall show that $f(x)$ is continuous at c. Let $\epsilon > 0$, and choose N such that
$$|f_N(x) - f(x)| < \frac{\epsilon}{3}$$
for all x in $[a,b]$. Since $f_N(x)$ is continuous, there is a $\delta > 0$ such that
$$|f_N(x) - f(x)| < \frac{\epsilon}{3}$$
whenever $|x - c| < \delta$. Hence
$$|f(x) - f(c)| \leq |f_N(x) - f(x)| + |f_N(x) - f_n(c)| + |f_N(c) - f(c)|$$
$$< \frac{\epsilon}{3} + \frac{\epsilon}{3} + \frac{\epsilon}{3} = \epsilon$$
when $|x - c| < \delta$. □

Example Consider the sequence of functions $f_n(x) = x^n$ on the interval $[0,1]$. We have

$$\lim x^n = \begin{cases} 0 & \text{if } 0 \le x < 1 \\ 1 & \text{if } x = 1. \end{cases}$$

The limit function is not continuous, so the convergence cannot be uniform. However, the convergence *is* uniform on any interval $[0, a]$ with $0 < a < 1$, and the function is continuous there. ∎

7.3.1 Problems

1. Let $f_n(x)$ be a sequence of continuous functions on $[0, 1]$, converging uniformly to $f(x)$. If x_n is a sequence in $[0, 1]$, with $\lim x_n = a$, prove that

   $$\lim f_n(x_n) = f(a).$$

   Show by a counterexample that this may fail if the convergence is not uniform.

2. Does the sequence

   $$f_n(x) = (1 - x^2)^n$$

   converge uniformly on $[-1, 1]$?

3. Discuss the convergence of the sequence

   $$f_n(x) = \frac{x^n - 1}{x^n + 1}$$

   on the real line $\mathbb{R}$. What is the limit? Is convergence uniform on $\mathbb{R}$? On $[-a, a]$ for a finite?

4. Let

   $$f_n(x) = \frac{2nx}{1 + n^2 x^2}.$$

   Prove that $f_n(x)$ converges to a continuous function, but that the limit is not uniform.

5. Let

   $$f_n(x) = \begin{cases} nx & 0 \le x \le 1/n \\ 2 - nx & 1/n \le x \le 2/n \\ 0 & \text{otherwise.} \end{cases}$$

   Prove that $f_n(x)$ converges to a continuous function, but that the limit is not uniform.

7.4 The Weierstrass M-Test

Let $f_n(x)$ be a sequence of functions on a set D. The series of functions

$$\sum_{n=1}^{\infty} f_n(x)$$

is said to *converge uniformly on D* to $f(x)$ iff the sequence

$$s_n(x) = \sum_{k=1}^{n} f_k(x)$$

of partial sums converges uniformly to $f(x)$ on D.

A standard condition that a series converge uniformly is the following.

Theorem 2 (Weierstrass M-Test) *Let $f_n(x)$ be a sequence of functions on a set D, and assume that there exists a sequence of numbers M_n such that*

$$|f_n(x)| \leq M_n$$

for all $x \in D$, and

$$\sum_{n=1}^{\infty} M_n < \infty. \qquad (7.3)$$

Then the series

$$\sum_{n=1}^{\infty} f_n(x)$$

converges uniformly on D.

Proof By the Comparison Test and (7.3) the series converges for each fixed point $x \in D$ (i.e., the series converges pointwise on D). Let $f(x)$ be the sum, so that

$$f(x) = \sum_{n=1}^{\infty} f_n(x)$$

for each $x \in D$. We need to show that the convergence is uniform.

We have

$$s_n(x) - f(x) = \sum_{k=n+1}^{\infty} f_k(x)$$

so that

$$|s_n(x) - f(x)| \leq \sum_{k=n+1}^{\infty} |f_k(x)| \leq \sum_{k=n+1}^{\infty} M_k.$$

Let $\epsilon > 0$, and choose N such that

$$\sum_{k=n+1}^{\infty} M_k < \epsilon$$

for $n \geq N$. Then for $n \geq N$, we have

$$|s_n(x) - f(x)| < \epsilon$$

for all $x \in D$. □

Example For the series

$$f(x) = \sum_{n=0}^{\infty} \frac{1}{n^2} \sin(nx)$$

we can take $M_n = 1/n^2$. The series converges uniformly since

$$\sum_{n=0}^{\infty} M_n = \sum_{n=0}^{\infty} \frac{1}{n^2} < \infty.$$

Thus, its sum $f(x)$ is a continuous function. ∎

7.4.1 Problems

1. Let

$$f(x) = \sum_{n=1}^{\infty} \frac{1}{1+n^2 x}.$$

For what set of values of x does this series converge? Is $f(x)$ continuous on this set? Is it bounded?

2. Let

$$H(x) = \begin{cases} 1 & \text{if } x \geq 0 \\ 0 & \text{if } x < 0. \end{cases}$$

Let a_n be any sequence of points, and define

$$f(x) = \sum_{n=1}^{\infty} c_n H(x - a_n)$$

where $c_n > 0$ and $\sum_{n=1}^{\infty} c_n < \infty$.

Prove that $f(x)$ is continuous at every point $x \neq a_n$.

3. Prove that the series

$$\sum_{n=1}^{\infty} (-1)^{n+1} \frac{x^2 + n}{n^2}$$

converges *uniformly* on every bounded interval, but converges *absolutely* for no value of x.

7.5 Integration

The integrals of a uniformly convergent sequence of functions converges to the integral of the limit.

Theorem 3 *Let $f_n(x)$ be integrable and converge to an integrable function $f(x)$ uniformly on the interval $[a, b]$. Then*

$$\lim \int_a^b f_n(x)\,dx = \int_a^b f(x)\,dx.$$

Proof Let $\epsilon > 0$, and choose N such that

$$|f_n(x) - f(x)| < \frac{\epsilon}{(b-a)}$$

for all x in $[a, b]$. Then

$$\left| \int_a^b f_n(x)\,dx - \int_a^b f(x)\,dx \right| = \left| \int_a^b f_n(x) - f(x)\,dx \right|$$

$$\leq \int_a^b |f_n(x) - f(x)|\,dx \leq \int_a^b \frac{\epsilon}{(b-a)}\,dx = \epsilon. \quad \square$$

The condition that $f(x)$ be integrable is unnecessary. The uniform limit of integrable functions is automatically integrable.

Theorem 4 *Let $f_n(x)$ be integrable and converge to $f(x)$ uniformly on the interval $[a, b]$. Then $f(x)$ is integrable on $[a, b]$.*

Proof Let $\epsilon > 0$ and let π be any partition of $[a, b]$. Choose N such that
$$|f(x) - f_N(x)| < \epsilon' = \frac{\epsilon}{4(b-a)}$$
for all x in $[a, b]$, and let
$$M_i(N) = \sup\{f_N(x) : x_{i-1} \le x \le x_i\}.$$
We must have
$$f_N(x) < f(x) + \epsilon' < M_i + \epsilon'$$
so that
$$M_i(N) \le M_i + \epsilon'$$
and by symmetry,
$$M_i \le M_i(N) + \epsilon'$$
so that
$$|M_i - M_i(N)| < \epsilon'.$$
It follows that
$$|U(f, \pi) - U(f_N, \pi)| = \left|\sum_{i=1}^n (M_i - M_i(N))\Delta x_i\right| < \sum_{i=1}^n \epsilon' \Delta x_i = \frac{\epsilon}{4}.$$
Similarly, if
$$m_i(N) = \inf\{f_N(x) : x_{i-1} \le x \le x_i\},$$
then
$$|m_i - m_i(N)| < \epsilon'$$
and
$$|L(f, \pi) - L(f_N, \pi)| < \frac{\epsilon}{4}.$$
Now choose π such that
$$U(f_N, \pi) - L(f_N, \pi) < \frac{\epsilon}{2}.$$
Then
$$|U(f, \pi) - L(f, \pi)|$$
$$\le |U(f, \pi) - U(f_N, \pi)| + |U(f_N, \pi) - L(f_N, \pi)| + |L(f_N, \pi) - L(f, \pi)|$$
$$< \frac{\epsilon}{4} + \frac{\epsilon}{2} + \frac{\epsilon}{4} = \epsilon.$$

Hence, $f(x)$ satisfies the Riemann Condition and is therefore integrable. □

7.5.1 Problems

1. Let
$$H(x) = \begin{cases} 1 & \text{if } x \geq 0 \\ 0 & \text{if } x < 0 \end{cases}$$

and define
$$f(x) = \sum_{n=1}^{\infty} \frac{1}{2^n} H(x-n).$$

Show that $f(x)$ is integrable and find
$$\int_0^a f(x)dx$$

where $a > 0$.

2. Let
$$f_n(x) = \frac{1}{\pi} \frac{n}{1+n^2 x^2}.$$

Find $\lim f_n(x)$ and
$$\lim \int_{-\infty}^{\infty} f_n(x)dx.$$

Draw a picture to see what is happening.

7.6 Differentiation

If the derivatives of a uniformly convergent sequence also converge, then the limit is the derivative of the limit function.

Theorem 5 *Let $f_n(x)$ be differentiable and $f'_n(x)$ integrable on $[a,b]$. If*

(a) *$f'_n(x)$ converges uniformly to a function $g(x)$ on $[a,b]$, and*

(b) *$f_n(c)$ converges for one fixed point c in $[a,b]$, then $f_n(x)$ converges uniformly on $[a,b]$ to a function $f(x)$ with*

$$f'(x) = g(x).$$

Proof By the Fundamental Theorem

$$f_n(x) = f_n(c) + \int_c^x f_n'(t)\, dt.$$

By Theorem 4, the function $f_n(x)$ converges to

$$f(x) = f(c) + \int_c^x g(t)\, dt$$

for each fixed x, which implies that $f'(x) = g(x)$. The convergence is uniform because, again by Theorem 4,

$$|f_n(x) - f(x)| = \left| f_n(c) - f(c) + \int_c^x f_n'(t) - g(t)\, dt \right|$$
$$\leq |f_n(c) - f(c)| + \int_a^b |f_n'(t) - g(t)|\, dt \to 0. \qquad \square$$

7.6.1 Problems

1. Prove that the function

$$\sum_{n=0}^\infty \frac{1}{n^3} \sin(nx)$$

   is continuously differentiable.

2. Prove that the function

$$\sum_{n=0}^\infty \frac{1}{2^n} \sin(nx)$$

   has derivatives of all orders.

7.7 *Iterated Limits

A limit of the form

$$\lim_{m \to \infty} \lim_{n \to \infty} a_{nm}$$

in which successive limits are taken is called an *iterated limit*. It is very important to realize that the result obtained may (or may not) *depend on the order in which the limits are taken.*

An example is the following:

$$\lim_{n\to\infty}\left(\lim_{m\to\infty}\frac{m}{n+m}\right) = \lim_{n\to\infty} 1 = 1$$

$$\lim_{m\to\infty}\left(\lim_{n\to\infty}\frac{m}{n+m}\right) = \lim_{m\to\infty} 0 = 0.$$

Double Limits

One condition under which the iterated limits are equal is the existence of the double limit.

Definition 2 The *double limit* $\lim_{n,m\to\infty} a_{nm}$ of a_{nm} is defined as follows. We say that

$$\lim_{n,m\to\infty} a_{nm} = L$$

iff for every $\epsilon > 0$, there exists an N such that $|a_{nm} - L| < \epsilon$ whenever both n and m are greater than N.

Theorem 6 If the double limit $\lim_{n,m\to\infty} a_{nm}$ exists, then both iterated limits exist and

$$\lim_{m\to\infty}\lim_{n\to\infty} a_{nm} = \lim_{n\to\infty}\lim_{m\to\infty} a_{nm}.$$

Proof Let $\lim_{n,m\to\infty} a_{nm} = L$ and let $|a_{nm} - L| < \epsilon$ for $n, m \geq N$. Then for a fixed $n > N$, we have $|a_{nm} - L| < \epsilon$, if $m > N$. This implies that

$$\left|\lim_{m\to\infty} a_{n,m} - L\right| \leq \epsilon$$

for all $n > N$. It follows also that

$$\left|\lim_{n\to\infty}\left(\lim_{m\to\infty} a_{n,m}\right) - L\right| \leq \epsilon.$$

Since ϵ is arbitrary, $\lim_{n\to\infty}(\lim_{m\to\infty} a_{n,m}) = L$. By symmetry, we have also

$$\lim_{m\to\infty}\left(\lim_{n\to\infty} a_{n,m}\right) = L. \qquad \square$$

Remark Questions of whether the order of two limiting processes can be exchanged to give the same answer are among the most important, and sometimes the most difficult, questions in analysis. In fact, since derivatives and integrals are actually limits, most of the questions considered in this chapter are of this type. For example, convergence of derivatives in (7.2) may be written as

$$\lim_{n\to\infty}\lim_{h\to 0}\frac{f_n(x+h)-f_n(x)}{h} = \lim_{h\to 0}\lim_{n\to\infty}\frac{f_n(x+h)-f_n(x)}{h}$$

and convergence of integrals as

$$\lim_{n\to\infty}\lim_{m\to 0}\sum_{k=1}^{n} f_m(\xi_i)\Delta x_i = \lim_{m\to 0}\lim_{n\to\infty}\sum_{k=1}^{n} f_m(\xi_i)\Delta x_i.$$

7.7.1 Problems

1. For which of the following double sequences are the two iterated limits equal?

   (a) $a_{nm} = \dfrac{mn}{n^2+m^2}$

   (b) $a_{nm} = \dfrac{mn+1}{n^2+nm}$

   (c) $a_{nm} = \dfrac{2mn}{(n+1)(m+2)}$

   (d) $a_{nm} = \dfrac{mn}{n^2+1}$

2. Prove that the double limit of

   $$\lim_{n,m\to\infty}\frac{2mn}{(n+1)(m+2)}$$

   exists.

3. Prove that the double limit of

   $$\lim_{n,m\to\infty}\frac{mn}{n^2+m^2}$$

   does not exist, although both iterated limits are equal.

4. Explain how the statement "$f(x)$ is continuous at a" asserts the interchange of two limits.

*5. Find the limit

   $$\lim_{m\to\infty}\lim_{n\to\infty}\cos^{2n}(m!\pi x).$$

7.8 Supplementary Problems

*1. Let $f_n(x)$ be a sequence of *monotone* functions on the closed bounded interval $[a, b]$ that converges pointwise to a *continuous* function $f(x)$. Prove that $f_n(x)$ converges to $f(x)$ uniformly.

Show by a counterexample that the theorem is false if $f(x)$ is not continuous.

(*Hint:* Assume first that the $f_n(x)$ are all increasing. The result follows from this case.)

*2. (*Dini's Theorem*) Let $f_n(x) \geq f_{n+1}(x) \geq \cdots \geq 0$ be a decreasing sequence of positive continuous functions on $[a, b]$ that converges pointwise to a *continuous* function $f(x)$. Prove that $f_n(x)$ converges to $f(x)$ uniformly.

Show by a counterexample that the theorem is false if $f(x)$ is not continuous.

(*Hint:* Replace $f_n(x)$ by $f_n(x) - f(x)$ and assume that $f_n(x) \downarrow 0$. Let the maximum of $f_n(x)$ occur at x_n and apply Bolzano–Weierstrass.)

3. Consider the series

$$f(x) = \sum_{n=0}^{\infty} \frac{x^2}{1+x^2} \left(\frac{1}{1+x^2}\right)^n.$$

(a) Prove that the series converges absolutely for all real x.
(b) Does the series converge uniformly on $[-1, 1]$?
(c) Is $f(x)$ continuous?

Chapter 8

Power Series

8.1 Power Series

An important example of a sequence of functions is a *power series*. A series of the form

$$\sum_{n=0}^{\infty} a_n (x-a)^n$$

is called a *power series about the point a*. The partial sum

$$s_n(x) = \sum_{k=0}^{n} a_n (x-a)^k$$

of a power series is a polynomial of degree n, which may converge to some function for some range of values of x.

Example 1 The *geometric series*

$$\frac{1}{1-x} = \sum_{n=0}^{\infty} x^n$$

converges to the function $1/(1-x)$ for $-1 < x < 1$. ∎

The basic fact about convergence of power series is the following.

Theorem 1 *If the terms of the series $\sum_{n=0}^{\infty} a_n c^n$ are bounded and $0 < r < |c|$, then $\sum_{n=0}^{\infty} a_n x^n$ is uniformly convergent on $|x| \leq r$.*

Proof Let $|a_n c^n| \leq M$. Then for $|x| \leq r$,

$$|a_n x^n| \leq |a_n r^n| \leq \left(\frac{r}{|c|}\right)^n |a_n c^n| \leq M \left(\frac{r}{|c|}\right)^n.$$

The result follows by the M-test. $\square$

It follows from this that the set of values for which a power series converges is always an interval centered at a (except when it consists only of the point a.)

Corollary 1 (Radius of Convergence) *There exists a number R, $0 \leq R \leq \infty$, such that the series*

$$\sum_{n=0}^{\infty} a_n (x-a)^n$$

is

(a) *uniformly and absolutely convergent on $|x| \leq r$ if $0 < r < R$, and*

(b) *divergent if $|x - a| > R$.*

Proof Take $a = 0$ for simplicity. Let

$$R = \sup \{c \geq 0 : a_n c^n \text{ is bounded.}\}$$

If $0 < r < R$, then by Theorem 1, $\sum_{n=0}^{\infty} a_n x^n$ converges uniformly and absolutely for $|x| \leq r$.

If $|x| > R$, then $a_n x^n$ is unbounded, so the series must diverge. $\square$

The number R is called the *Radius of Convergence* of the series, and the interval $(a - R, a + R)$ is called its *Interval of Convergence*.

Example 2 We know that the geometric series converges on the interval $-1 < x < 1$. Thus, here, $R = 1$ and the Interval of Convergence is $-1 < x < 1$.

Note that we may have $R = \infty$, as for the exponential series

$$\sum_{n=0}^{\infty} \frac{x^n}{n!},$$

which converges at all points. On the other hand, the series

$$\sum_{n=0}^{\infty} n! x^n$$

has $R = 0$, and converges only for $x = a$. ∎

Theorem 1 says nothing about what happens at the *endpoints* of the Interval of Convergence. More or less anything can happen there, as Problem 3 shows.

The *Ratio Test* is frequently useful in finding the Radius of Convergence.

Example 3 Consider the series

$$\sum_{n=0}^{\infty} \frac{n^2}{2^n} (x-2)^n.$$

According to the Ratio Test, the series is convergent if

$$\lim \frac{\left|\frac{(n+1)^2}{2^{n+1}} (x-2)^{n+1}\right|}{\left|\frac{n^2}{2^n} (x-2)^n\right|} = \lim \frac{(n+1)^2}{2n^2} |x-2| = \frac{|x-2|}{2} < 1,$$

and divergent if

$$\frac{|x-2|}{2} > 1.$$

The Interval of Convergence is therefore $|x-2| < 2$. ∎

One can also use the *Root Test*. For the same series, this gives for convergence

$$\lim \left|\frac{n^2}{2^n} (x-2)^n\right|^{1/n} = \frac{|(x-2)|}{2} \lim n^{2/n} = \frac{|(x-2)|}{2} < 1.$$

8.1.1 *Hadamard's Formula

A generalization of the Root Test will always—in principle—give the Radius of Convergence.

Theorem 2 (Hadamard's Formula) *The Radius of Convergence R of $\sum_{n=0}^{\infty} a_n (x-a)^n$ is given by*

$$\frac{1}{R} = \limsup |a_n|^{1/n}. \tag{8.1}$$

Proof Take $a=0$ for simplicity, and let R be defined by (8.1). Let $|x| < R$, and choose ρ with $|x| < \rho < R$. Then, since

$$\frac{1}{\rho} > \frac{1}{R}$$

there exists an N such that, for $n \geq N$,

$$|a_n|^{1/n} < \frac{1}{\rho}.$$

Hence, if $|x| \leq r$, we have for $n \geq N$,

$$|x|^n |a_n| \leq r^n |a_n| < \left(\frac{r}{\rho}\right)^n$$

so that the series converges uniformly by the Weierstrass M-test.

On the other hand, let $|x| > R$, and choose ρ with $|x| > \rho > R$. Since

$$\frac{1}{\rho} < \frac{1}{R}$$

there must exist a subsequence a_{n_k} of a_n such that

$$|a_{n_k}|^{1/n} > \frac{1}{\rho}.$$

Hence, if $|x| > \rho$, we have

$$|x|^{n_k} |a_{n_k}| > \left(\frac{r}{\rho}\right)^{n_k} \to \infty.$$

The series cannot converge because its nth term does not tend to zero. □

8.1.2 Problems

1. Find the interval of convergence of these series.

(a) $\sum_{n=0}^{\infty} n^2 x^n$

(b) $\sum_{n=1}^{\infty} \frac{2^n}{n^2} (x-1)^n$

(c) $\sum_{n=0}^{\infty} (-1)^n \frac{1}{(2n+1)!} x^n$

(d) $\sum_{n=0}^{\infty} \frac{n!}{2^n} x^n$

(e) $\sum_{n=1}^{\infty} \left(\frac{n+1}{n}\right)^{n^2} x^n$

(f) $\sum_{n=0}^{\infty} x^{n!}$

2. Series of functions that are not power series may have regions of convergence that are not intervals. On what regions do the following series converge?

(a) $\sum_{n=0}^{\infty} n \left(\frac{2}{x}\right)^n$

(b) $\sum_{n=1}^{\infty} \frac{(x(x-1))^n}{2^n}$

(c) $\sum_{n=0}^{\infty} \frac{1}{n^2} \left(\frac{4x}{1+x^2}\right)^n$

(d) $\sum_{n=1}^{\infty} 2^n \sin^n x$

3. Theorem 1 says nothing about what happens at the endpoints of the Interval of Convergence. As examples of what can happen, examine the convergence at endpoints of the following series. Distinguish between conditional and absolute convergence.

(a) $\sum_{n=1}^{\infty} (-1)^n \frac{1}{n} x^{2n}$

(b) $\sum_{n=1}^{\infty} \frac{1}{n^2} x^n$

(c) $\sum_{n=1}^{\infty} \frac{1}{n} x^n$

(d) $\sum_{n=1}^{\infty} \frac{1}{n} x^n$

4. According to Problem 3 of Section 2.10, the Root Test will always give the Radius of Convergence whenever the Ratio Test does. Show, however, that the Root Test may sometimes give the Radius of Convergence when the Ratio Test fails. (*Hint:* Try $\sum_{n=1}^{\infty} e^{(-1)^n \sqrt{n}} x^n$.)

8.2 Operations on Power Series

A power series may be integrated and differentiated term-by-term inside its Interval of Convergence.

Theorem 3 (Integration) *Let R be the Radius of Convergence of the series*

$$f(x) = \sum_{n=0}^{\infty} a_n x^n.$$

Then

$$\int_0^x f(t)dt = \sum_{n=0}^{\infty} \frac{a_n}{n+1} x^{n+1}$$

for $|x| < R$.

Proof Fix x with $|x| < R$. The sequence of partial sums

$$s_n(t) = \sum_{k=0}^{n} a_k t^k$$

converges uniformly on $|t| \leq |x|$. Hence, by Theorem 4 of Chapter 7,

$$\int_0^x s_n(t)dt = \int_0^x \sum_{k=0}^{n} a_k t^k dt = \sum_{k=0}^{n} a_k \frac{1}{k+1} x^{k+1}$$

converges to

$$\int_0^x f(t)dt. \qquad \square$$

Example 1 Integrating the *geometric series* gives

$$-\log(1-x) = \int_0^x \frac{1}{1-t} dt = \sum_{n=0}^{\infty} \int_0^x t^n dt = \sum_{n=0}^{\infty} \frac{1}{n+1} x^{n+1}$$

$$= x + \frac{x^2}{2} + \frac{x^3}{3} + \cdots$$

for $-1 < x < 1$. ∎

Theorem 4 (Differentiation) *Let R be the radius of convergence of the series*

$$f(x) = \sum_{n=0}^{\infty} a_n x^n.$$

Then

$$f'(x) = \sum_{n=0}^{\infty} na_n x^n$$

for $|x| < R$.

Proof We only need to show that the Radius of Convergence of the differentiated series is equal to R. For then the sequence

$$s'_n(x) = \sum_{k=1}^{n} ka_k x^{k-1}$$

of derivatives of partial sums is uniformly convergent on $|x| \leq r$ for $0 < r < R$, and so, by Theorem 5 of Chapter 7,

$$f'(x) = \lim s'_n(x) = \sum_{n=0}^{\infty} na_n x^n.$$

There are two ways prove that $s'_n(x)$ converges uniformly. If we let $0 < r < \rho < R$, and $|x| \leq r$, then, by the estimate in the proof of Theorem 1,

$$|na_n x^n| \leq nr^n |a_n| \leq n\left(\frac{r}{\rho}\right)^n$$

for $n \geq N$. But

$$\sum_{n=0}^{\infty} n\left(\frac{r}{\rho}\right)^n < \infty.$$

For the second proof, note that Hadamard's formula states that the Radius of Convergence of the differentiated series is

$$\limsup (n|a_n|)^{1/n} = \limsup n^{1/n} |a_n|^{1/n} = \left(\lim n^{1/n}\right) \limsup |a_n|^{1/n} = R$$

(by Problem 4, Section 2.9), since

$$\lim n^{1/n} = 1. \qquad \square$$

Example 2 Differentiating the geometric series

$$\frac{1}{1-x} = \sum_{k=0}^{\infty} x^k = 1 + x + x^2 + x^3 + \cdots$$

gives

$$\frac{1}{(1-x)^2} = \frac{d}{dx}\frac{1}{1-x} = \sum_{k=0}^{\infty} \frac{d}{dx} x^k$$

$$= \sum_{k=0}^{\infty} k x^k = 1 + 2x + 3x^2 + 4x^3 + \cdots.$$

This formula is valid on the Interval of Convergence of the original series, $-1 < x < 1$. ∎

8.2.1 Problems

1. Find the series expansions of these functions about the point $a = 0$.

   (a) $\dfrac{1}{(4+x^2)^3}$

   (b) $\dfrac{1}{2} \log\left(\dfrac{1+x}{1-x}\right)$

   (c) $\arctan(x)$

   (*Hint for* (c): What is its derivative?)

2. Sum the following series.

   (a) $\displaystyle\sum_{n=0}^{\infty} \frac{2^n}{3^n}$

   (b) $\displaystyle\sum_{n=1}^{\infty} (-1)^{n+1} \frac{n}{3^n}$

   (c) $\dfrac{1}{1 \cdot 2^1} + \dfrac{1}{2 \cdot 2^2} + \dfrac{1}{3 \cdot 2^3} + \dfrac{1}{4 \cdot 2^4} + \cdots$

8.3 Taylor's Theorem

One consequence of the differentiation theorem is the familiar formula for the coefficients in a power series.

8.3 Taylor's Theorem

Theorem 5 (Taylor's Formula) *If*

$$f(x) = \sum_{n=0}^{\infty} a_n (x-a)^n$$

then

$$a_n = \frac{f^{(n)}(a)}{n!}.$$

Proof Take $a = 0$ for simplicity. By Theorem 4, applied n times, the nth derivative of

$$f(x) = \sum_{n=0}^{\infty} a_n x^n$$

is

$$f^{(n)}(x) = \sum_{k=n}^{\infty} a_k k(k-1) \cdots (k-n+1) x^{k-n}.$$

Set $x = 0$. All terms but the first drop out, and we obtain

$$f^{(n)}(0) = a_n n(n-1) \cdots 2 \cdot 1 = a_n n!.$$ □

Example 1 Consider the function $f(x) = e^x$. Since this function is its own derivative, we have $f^{(n)}(x) = e^x$ for all $n \geq 0$. Hence, $f^{(n)}(0) = 1$, and Taylor's formula gives

$$a_n = \frac{1}{n!}$$

for the coefficient a_n. Therefore, *if the function e^x has a power series expansion about $x = 0$*, it must be given by the familiar formula

$$e^x = \sum_{n=0}^{\infty} \frac{1}{n!} x^n.$$ ■

It is important to realize, however, that *we have not yet proved that this formula holds*. We do know that

(a) If there is such an expansion, this must be it, and

(b) By the Ratio Test, the series on the right side converges.

What we do not yet know is whether the sum of this series is *actually equal to* the function e^x. It is entirely possible for a function to be infinitely differentiable, in which case the Taylor coefficients can be computed, and yet, either

(a) The series so computed converges to a different function, or

(b) The series is divergent.

Example 2 As an example of (a), consider the function
$$f(x) = e^{-1/x^2}$$
with $f(0) = 0$. For $x \neq 0$, its derivative is
$$f'(x) = \frac{2}{x^3} e^{-1/x^2}.$$
At $x = 0$, the derivative is
$$f'(0) = \lim_{x \to 0} \frac{f(x) - f(0)}{x} = \lim_{x \to 0} \frac{1}{x} e^{-1/x^2} = 0.$$
By induction, we may prove that for $x \neq 0$,
$$f^{(n)}(x) = Q_n(1/x) e^{-1/x^2}$$
and $f^{(n)}(0) = 0$ for all $n \geq 0$, where $Q_n(1/x)$ is a polynomial in $1/x$. It follows that the Taylor series for $f(x)$ has coefficients
$$\frac{f^{(n)}(0)}{n!} = 0$$
for all n. The Taylor series for e^{-1/x^2} therefore converges to zero, not e^{-1/x^2}. ∎

An example of (b) is given in Problem 5.

Taylor's Theorem with Remainder

In order to prove that
$$e^x = \sum_{n=0}^{\infty} \frac{1}{n!} x^n$$

really converges to e^x, we need to *estimate the difference between the partial sum*

$$s_n(x) = \sum_{k=0}^{n} \frac{1}{k!} x^k$$

and the function e^x. This is done by *Taylor's formula with remainder*.

Definition 1 Let $f(x)$ be an n times differentiable function on an interval containing the point a. The *nth Taylor polynomial $P_n(x)$ at a*, is defined by

$$P_n(x) = \sum_{k=1}^{n} \frac{f^{(k)}(a)}{k!} (x-a)^k.$$

The Taylor polynomial is just the *nth* partial sum of the power series for $f(x)$ at $x = a$. However, the Taylor polynomial will exist for any function that is n times differentiable at $x = a$. It is the unique polynomial of degree n that has the same derivatives as the function $f(x)$ at $x = a$.

Lemma 1 *The values of $f(x)$ and $P_n(x)$ and their first n derivatives at a agree; that is,*

$$P_n(a) = f(a), \ P_n'(a) = f'(a), \ldots, P_n^{(n)}(a) = f^{(n)}(a).$$

Theorem 6 (Taylor's Theorem with Lagrange Remainder) *Let $f(x)$ be $n+1$ times differentiable on an interval $(a - \delta, a + \delta)$. If x is in $(a - \delta, a + \delta)$, then there exists a point c, between a and x, with*

$$f(x) = \sum_{k=1}^{n} \frac{f^{(k)}(a)}{k!}(x-a)^k + R_n(x; a, f)$$

where

$$R_n(x; a, f) = \frac{f^{(n+1)}(c)}{(n+1)!}(x-a)^{n+1}.$$

This formula for the remainder R_n is called the *Lagrange form* of the remainder. There are other forms of R_n, one of which is given in the next chapter.

If $n = 0$, then $P_0(a) = f(a)$, and Taylor's formula reduces to

$$f(x) = f(a) + f'(c)(x-a),$$

which is the Mean Value Theorem. *Taylor's Theorem is therefore a generalization of the Mean Value Theorem.*

The Mean Value Theorem was proved using Rolle's Theorem, and we will use a generalization of Rolle's Theorem to prove Taylor's Theorem.

Theorem 7 (Generalized Rolle's Theorem) *Let $g(x)$ be n times differentiable on $a < x < b$ and assume that $g^{(k)}(x)$ is continuous on $a \leq x \leq b$ for $k = 1, ..., n$. If*

$$g(a) = g'(a) = g''(a) = \ldots = g^{(n-1)}(a) = g(b) = 0,$$

then there exists a point c, $a < c < b$, with

$$g^{(n)}(c) = 0.$$

Proof of Theorem 7 Since $g(a) = g(b) = 0$, by Rolle's Theorem, there exists a point c_1 between a and b, such that $g'(c_1) = 0$. But now, since $g'(a) = g'(c_1) = 0$, there exists a point c_2 between a and c_1, such that $g''(c_2) = 0$. Continuing in this manner, we find a point $c = c_n$ where $g^{(n)}(c) = 0$. □

Proof of Taylor's Theorem Fix x, and let

$$g(t) = f(t) - P_n(t) - M(t-a)^{n+1}$$

where M is chosen so that $g(x) = 0$. The function $g(t)$ then satisfies the hypotheses of Theorem 7 on the interval $[0, x]$, so there exists c between a and x such that

$$g^{(n+1)}(c) = 0.$$

But since $P_n(x)$ is a polynomial of degree n, its $(n+1)$th derivative is identically zero. Hence,

$$g^{(n+1)}(c) = f^{(n+1)}(c) - M(n+1)! = 0$$

so that

$$M = \frac{f^{(n+1)}(c)}{(n+1)!}.$$

Hence, setting $t = x$ gives

$$f(x) - P_n(x) - \frac{f^{(n+1)}(c)}{(n+1)!}(x-a)^{n+1} = g(x) = 0. \quad \square$$

Example 3 We are now able to show that

$$\sum_{n=0}^{\infty} \frac{x^n}{n!}$$

actually converges to e^x. For since $f^{(n)}(x) = e^x$, Taylor's theorem gives

$$e^x = \sum_{k=0}^{n} \frac{x^k}{k!} + R_n$$

where

$$R_n = \frac{x^{n+1}}{(n+1)!} e^c$$

for some c between 0 and x. But then

$$0 < R_n < e^x \frac{x^n}{(n+1)!} < e^x \frac{x^n}{(n+1)!} \to 0.$$ ∎

The Meaning and Application of Power Series

This is a good place to comment on the significance and usefulness of power series. The key is Lemma 1. It says that, taking $a = 0$, the Taylor polynomial, which is the *nth* partial sum of the series, has the same derivatives as the function $f(x)$ at the origin. This means that the graph of the Taylor polynomial sticks very close to the graph of $f(x)$ near the origin. Therefore, *$P_n(x)$ is a good approximation to $f(x)$ for small x.*

For this reason, power series are useful in calculating various functions numerically. The remainder formula in Taylor's Theorem lets us estimate the error made in using $P_n(x)$ to calculate $f(x)$.

This explains what is going on in Example 2. The function e^{-1/x^2} goes to zero as x tends to zero *faster than any power of x*, and so cannot be represented well near zero by any polynomial except zero.

Example 4 As an illustration, let us use the formula

$$e = \sum_{n=0}^{\infty} \frac{1}{n!} = 1 + 1 + \frac{1}{2!} + \ldots + \frac{1}{n!} + R_n$$

to compute the number e to five decimal places. Using that $e < 3$, we have for R_n, with $0 < c < 1$,

$$0 < R_n = \frac{e^c}{(n+1)!} < \frac{e}{(n+1)!} < \frac{3}{(n+1)!}.$$

Taking $n = 9$,

$$0 < e - \left(1 + 1 + \frac{1}{2!} + \ldots + \frac{1}{9!}\right) < R_9 < \frac{3}{10!} < 0.83 \times 10^{-6} < 0.000001.$$

We compute that
$$\left(1 + 1 + \frac{1}{2!} + \ldots + \frac{1}{9!}\right) = 2.7182815.$$
Thus,
$$2.7182815 < e < 2.7182825$$
so that, to five places,
$$e = 2.71828. \qquad \blacksquare$$

Irrationality of e

We can also use this series to prove that e is irrational.

Theorem 8 *e is irrational.*

Proof Suppose that e is rational; that is, $e = p/q$ where p and q are integers. Let $n > q$ and $n > 2$. Then
$$n!e - n!\left(1 + 1 + \frac{1}{2!} + \ldots + \frac{1}{n!}\right) = n!R_n.$$
The left side is an *integer*, while, for some c between 0 and 1,
$$0 < n!R_n = n!e^c \frac{1}{(n+1)!} < \frac{e}{(n+1)!} < \frac{3}{n+1} < 1.$$
This is a contradiction, since there are no integers between 0 and 1. $\qquad \square$

8.3.1 Problems

1. Prove Lemma 1.

2. Find the Taylor series about $a = 0$ for $\sin x$ and prove that it converges to $\sin x$. Use it to compute $\sin(1)$ to three decimal places.

3. Prove that the Taylor polynomial is the unique polynomial of degree n that has the same derivatives as the function $f(x)$ at $x = a$.

4. Prove by induction, that if $f(x) = e^{-1/x^2}$, then
$$f^{(n)}(x) = Q_n(1/x)e^{-1/x^2}$$
for $x \neq 0$, where $Q_n(1/x)$ is a polynomial in $1/x$, and therefore $f^{(n)}(0) = 0$ for all $n \geq 0$.

*5. Let
$$f(x) = \sum_{n=1}^{\infty} \frac{\cos(2^n x)}{n!}.$$

(a) Prove that $f(x)$ is infinitely differentiable.

(b) Find a formula for the nth derivative of $f(x)$. Show that
$$f^{(2k+1)}(0) = 0$$
and
$$f^{(2k)}(0) = (-1)^k \left[e^{4^k} - 1\right].$$

(c) Prove that the Taylor expansion of $f(x)$ about $x = 0$ diverges.

8.4 Supplementary Problems

*1. Find the radius of convergence of
$$\sum_{n=1}^{\infty} \sin n \; x^n.$$

(*Hint:* Use Theorem 1 directly.) Hence, find
$$\limsup (\sin n)^{1/n}.$$

2. Show that
$$\int_0^1 \frac{\log(1-x)}{x} dx = -\sum_{n=1}^{\infty} \frac{1}{n^2}.$$

Chapter 9
*Further Topics in Series

9.1 *Summation by Parts

Abel's Summation by Parts formula is the discrete analog of Integration by Parts. Let a_n be any sequence, and

$$s_n = \sum_{k=0}^{n} a_k$$

be the partial sums of the series

$$\sum_{k=0}^{\infty} a_k.$$

Set $s_{-1} = 0$.

Theorem 1 (Summation by Parts) *Let a_n and c_n be any sequences. Then*

$$\sum_{k=0}^{n} a_k c_k = s_n c_{n+1} + \sum_{k=0}^{n} s_k (c_k - c_{k+1}).$$

Proof Since $a_k = (s_k - s_{k-1})$, we have

$$\sum_{k=0}^{n} a_k c_k - \sum_{k=0}^{n} s_k (c_k - c_{k+1}) = \sum_{k=0}^{n} (s_k - s_{k-1}) c_k - s_k (c_k - c_{k+1})$$

$$= \sum_{k=0}^{n} (s_k c_k - s_{k-1} c_k - s_k c_k + s_k c_{k+1})$$

$$= \sum_{k=0}^{n} (s_k c_{k+1} - s_{k-1} c_k).$$

This sum telescopes, and is equal to

$$s_n c_{n+1} - s_{-1} c_0 = s_n c_{n+1}.$$ □

One application of Summation by Parts is Dirichlet's Test.

Theorem 2 (Dirichlet's Test) *Let $c_n \geq c_{n+1} > 0$ be a positive decreasing sequence, and assume that the partial sums s_n are bounded.*
   If either

(a) $\lim c_n = 0$, *or*

(b) $\sum_{n=0}^{\infty} a_n$ *converges,*

then $\sum_{k=0}^{\infty} a_k c_k$ converges.

Proof Let $|s_n| \leq M$. Let $c = \lim c_n$, which exists because c_n is a decreasing sequence bounded below. Summing by parts gives

$$\sum_{k=0}^{n} a_k c_k = s_n c_{n+1} + \sum_{k=0}^{n} s_k (c_k - c_{k+1}).$$

Now the series

$$\sum_{k=0}^{\infty} s_k (c_k - c_{k+1})$$

is absolutely convergent, since

$$|s_k (c_k - c_{k+1})| \leq M (c_k - c_{k+1}).$$

The sum of the right side telescopes to give

$$\sum_{k=0}^{n} (c_k - c_{k+1}) = c_0 - c_{n+1} \to c_0 - c.$$

It remains to show that $\lim s_n c_{n+1}$ exists. In case (a),

$$|s_n c_{n+1}| \leq M c_{n+1} \to 0$$

so that $\lim s_n c_{n+1} = 0$. In case (b), $\lim s_n = s$ exists and so $\lim s_n c_{n+1} = sc$. □

Example Take $a_n = (-1)^n$ and c_n decreasing to 0. In case (a), we get Leibnitz's Theorem. ∎

9.2 *The Theorems of Abel and Tauber

By Leibnitz's Test, the series

$$\sum_{n=0}^{\infty} \frac{(-1)^n}{n+1} = 1 - \frac{1}{2} + \frac{1}{3} - \frac{1}{4} + \cdots$$

is convergent. What is its value? For $-1 < x < 1$, we have

$$\log(1+x) = \sum_{n=0}^{\infty} \frac{1}{n+1} x^{n+1}.$$

If we could set $x = 1$, we would have

$$\sum_{n=0}^{\infty} \frac{(-1)^n}{n+1} = \log 2.$$

That this is permissible follows from the next theorem.

Theorem 3 (Abel) *Let*

$$f(x) = \sum_{n=0}^{\infty} a_n x^n$$

be convergent for $|x| < 1$. If $\sum_{n=0}^{\infty} a_n$ is convergent to s, then

$$s = f(1-) = \lim_{x \to 1-} f(x).$$

Proof We have, with $s_{-1} = 0$,

$$f(x) = \sum_{n=0}^{\infty} a_n x^n = \sum_{n=0}^{\infty} (s_n - s_{n-1}) x^n$$

$$= \sum_{n=0}^{\infty} s_n \left(x^n - x^{n+1} \right) = (1-x) \sum_{n=0}^{\infty} s_n x^n.$$

Let $\epsilon > 0$, and choose N such that $|s_n - s| < \epsilon$ for $n \geq N$. Since

$$(1-x) \sum_{n=0}^{\infty} x^n = 1$$

we have

$$f(x) - s = (1-x) \sum_{n=0}^{\infty} (s_n - s) x^n$$

$$= (1-x) \sum_{n=0}^{N} (s_n - s) x^n + (1-x) \sum_{n=N+1}^{\infty} (s_n - s) x^n.$$

But s_n is bounded. If we take $|s_n| \leq M$, then

$$|f(x) - s| \leq \left|(1-x)\sum_{n=0}^{N}(s_n - s)x^n\right| + \left|(1-x)\sum_{n=N+1}^{\infty}(s_n - s)x^n\right|$$

$$\leq 2MN\,|(1-x)| + \epsilon\left|(1-x)\sum_{n=N+1}^{\infty}x^n\right|$$

$$\leq 2MN\,|(1-x)| + \epsilon.$$

Hence, for every $\epsilon > 0$,

$$\limsup_{x \to 1-} |f(x) - s| \leq \epsilon$$

so that

$$\lim_{x \to 1-} f(x) = s. \qquad \square$$

The converse of Abel's Theorem would state that if $s = f(1-)$ exists, then $\sum_{n=0}^{\infty} a_n$ is convergent to s. However, this assertion is false. For example, if $a_n = (-1)^n$, then

$$f(x) = \sum_{n=0}^{\infty}(-1)^n x^n = \frac{1}{1+x}$$

so that $f(1-) = \frac{1}{2}$, while $\sum_{n=0}^{\infty}(-1)^n$ is divergent.

Tauber's Theorem is a partial converse to *Abel's Theorem*.

Theorem 4 (Tauber) *If $s = f(1-)$ exists, and if*

$$\lim na_n = 0,$$

then $\sum_{n=0}^{\infty} a_n$ is convergent to s.

Proof Write

$$s_n - s = f(x) - s + \sum_{k=0}^{n} a_k\left(1 - x^k\right) - \sum_{k=n+1}^{\infty} a_k x^k.$$

By Theorem 13 of Chapter 2, $\lim na_n = 0$ implies that the Cesaro limit

$$\lim \frac{1}{n}\sum_{k=1}^{n} k\,|a_k| = 0.$$

Let $x = 1 - \frac{1}{n}$; or, that is, $1 - x = 1/n$. Note that then

$$\sum_{j=0}^{\infty} x^j = \left(1 - \left(1 - \frac{1}{n}\right)\right)^{-1} = n$$

and hence

$$1 - x^k = (1-x)(1 + x + x^2 + \ldots + x^{k-1}) \leq k(1-x) = k/n. \tag{9.1}$$

Choose N so that for $n \geq N$,

$$\left| f(1 - \frac{1}{n}) - s \right| < \frac{\epsilon}{3}$$

$$|na_n| < \frac{\epsilon}{3}$$

and,

$$\frac{1}{n} \sum_{k=1}^{n} k |a_k| < \frac{\epsilon}{3}.$$

Since, for $0 \leq x < 1$, we have

$$\left| \sum_{k=0}^{n} a_k (1 - x^k) \right| \leq \frac{1}{n} \sum_{k=1}^{n} k |a_k| < \frac{\epsilon}{3}.$$

We then have, using 9.1,

$$\left| \sum_{k=n+1}^{\infty} a_k x^k \right| = \left| \sum_{j=0}^{\infty} a_{j+n+1} x^j x^{n+1} \right|$$

$$\leq x^{n+1} \sum_{j=0}^{\infty} \frac{\epsilon}{3(j+n+1)} x^j$$

$$\leq \frac{\epsilon}{3n} \sum_{j=0}^{\infty} x^j = \frac{\epsilon}{3}.$$

Combining these estimates then yields

$$|s_n - s| \leq \left| f(1 - \frac{1}{n}) - s \right| + \left| \sum_{k=0}^{n} a_k (1 - x^k) \right| + \left| \sum_{k=n+1}^{\infty} a_k x^k \right|$$

$$\leq \frac{\epsilon}{3} + \frac{\epsilon}{3} + \frac{\epsilon}{3} = \epsilon. \qquad \square$$

Any partial converse to Abel's Theorem is called a *Tauberian Theorem*. Here is another one.

Theorem 5 *If $L = f(1-)$ exists, and if $a_n \geq 0$, then $\sum_{n=0}^{\infty} a_n$ is convergent to L.*

Proof Let $s_n(x) = \sum_{k=0}^{n} a_k x^k$ and $s_n = \sum_{k=0}^{n} a_k$. Then, for $0 \leq x < 1$,

$$s_n(x) \leq f(x) \leq L.$$

Let $x \to 1$ to get

$$s_n = s_n(1) \leq L.$$

Thus s_n is a bounded increasing sequence, so $s = \lim s_n$ exists and $s \leq L$. Moreover,

$$f(x) = \sum_{k=0}^{\infty} a_k x^k \leq \sum_{k=0}^{\infty} a_k = s.$$

Let $x \to 1$ to get

$$L = f(1-) \leq s. \qquad \square$$

9.3 *The Integral Form of Taylor's Theorem

The following version of Taylor's Theorem expresses the remainder as an integral.

Theorem 6 (Taylor's Theorem with Integral Remainder) *Let $f(x)$ be $n+1$ times differentiable on an interval $(a-\delta, a+\delta)$. If x is in $(a-\delta, a+\delta)$, then*

$$f(x) = \sum_{k=1}^{n} \frac{f^{(k)}(a)}{k!}(x-a)^k + R_n(x; a, f)$$

where

$$R_n(x; a, f) = \frac{1}{n!} \int_a^x (x-t)^n f^{(n+1)}(t) dt. \qquad (9.2)$$

Proof Define $R_n(x; a, f)$ by 9.2. Then

$$R_0(x; a, f) = \int_a^x f^{(1)}(t) dt = f(x) - f(a).$$

If $f^{(n+2)}(t)$ exists, then, integrating by parts,

$$R_n(x;a,f) = \frac{1}{n!}\int_a^x (x-t)^n f^{(n+1)}(t)dt$$

$$= \frac{1}{n!}\left[\frac{-(x-t)^{n+1}}{n+1}f^{(n+1)}(t)\right]_a^x - \frac{1}{n!}\int_a^x \frac{-(x-t)^{n+1}}{n+1}f^{(n+2)}(t)dt$$

$$= \frac{1}{n!}f^{(n+1)}(a) + R_{n+1}(x;a,f).$$

The result follows by induction. □

A form of R_n that is sometimes more convenient may be obtained by taking $a = 0$ and letting $t = x(1-s)$, where $0 \leq s \leq 1$. Then $x - t = xs$ and $dt = -xds$, and we obtain

$$R_n(x) = \frac{1}{n!}\int_0^1 (xs)^n f^{(n+1)}(x(1-s))xds$$

$$= \frac{x^{n+1}}{n!}\int_0^1 s^n f^{(n+1)}(x(1-s))ds.$$

An application of the Integral form of Taylor's Theorem is the following result of S. Bernstein.

Theorem 7 (S. Bernstein) *Let $f(x)$ be infinitely differentiable, and suppose that $f^{(n)}(x) \geq 0$ for $0 \leq x \leq r$. Then the Taylor series of $f(x)$ converges to $f(x)$ for $0 < x < r$.*

Proof First note that since $f^{(k)}(0) \geq 0$, we have

$$R_n(r) \leq \sum_{k=1}^n \frac{f^{(k)}(0)}{k!}r^k + R_n(r) = f(r).$$

Moreover, since $f^{(n+1)}(t)$ is increasing on $0 \leq t \leq r$, we have

$$f^{(n+1)}(x(1-s)) \leq f^{(n+1)}(r(1-s))$$

for $0 \leq x \leq r$. Hence, by Theorem 6,

$$R_n(x) = \frac{x^{n+1}}{n!}\int_0^1 s^n f^{(n+1)}(x(1-s))ds$$

$$\leq \frac{x^{n+1}}{n!}\int_0^1 s^n f^{(n+1)}(r(1-s))ds$$

$$\leq \left(\frac{x}{r}\right)^{n+1}\frac{r^{n+1}}{n!}\int_0^1 s^n f^{(n+1)}(r(1-s))d$$

$$= \left(\frac{x}{r}\right)^{n+1} R_n(r) \leq \left(\frac{x}{r}\right)^{n+1} f(r).$$

Hence $R_n(x) \to 0$ for $0 < x < r$. □

Bernstein's Theorem provides a proof of Newton's Binomial Theorem (see Problem 3).

9.3.1 Problems

1. Use the Mean Value Theorem for Integrals to derive the Lagrange form of the remainder from the Integral form.

2. Prove that the function $f(x) = 1 - \sqrt{1-x}$ satisfies the hypotheses of Bernstein's Theorem.

3. (*Newton's Binomial Series*) Show that for any real α, the power series about $x = 0$ of $f(x) = (1-x)^\alpha$ converges for $|x| < 1$.

9.4 *Kummer's Test

Kummer's Test is a very general test for convergence of series of positive terms, which includes the Ratio Test as a special case. I have included it because its proof, which is remarkably brief, looks like black magic.

Theorem 8 (Kummer's Test) *Let $a_n > 0$. Suppose that there exists a positive sequence $c_n > 0$ and a positive number r such that*

$$\frac{a_n}{a_{n+1}} c_n - c_{n+1} > r. \tag{9.3}$$

Then $\sum_{n=1}^{\infty} a_n$ converges.
On the other hand, if

$$\frac{a_n}{a_{n+1}} c_n - c_{n+1} < 0 \tag{9.4}$$

and

$$\sum_{n=1}^{\infty} \frac{1}{c_n} = \infty,$$

then $\sum_{n=0}^{\infty} a_n$ diverges.

Proof Let $s_n = a_1 + a_2 + \cdots + a_n$. If (9.3) holds, we have

$$a_k c_k - c_{k+1} a_{k+1} \geq r a_{k+1} > 0.$$

Summing this gives

$$r(s_n - a_1) \le \sum_{k=1}^{n-1}(a_k c_k - c_{k+1}a_{k+1}) = a_1 c_1 - c_n a_n \le a_1 c_1$$

since the sum telescopes. Hence

$$s_n \le \frac{a_1 c_1 + r a_1}{r}$$

so that the partial sums are bounded.

If, on the other hand, (9.4) holds, then

$$a_n c_n < a_{n+1} c_{n+1}$$

so that

$$a_1 c_1 < a_2 c_2 < \cdots < a_n c_n$$

and

$$a_n > (a_1 c_1)\frac{1}{c_n}.$$

Thus

$$\sum_{n=1}^{\infty} a_n > (a_1 c_1) \sum_{n=1}^{\infty} \frac{1}{c_n} = \infty.$$

□

Note that if $c_n = 1$, Kummer's Test reduces to the Ratio Test.

One consequence of Kummer's Test is *Raabe's Test*.

Theorem 9 (Raabe's Test) *Let $a_n > 0$. Then $\sum_{n=1}^{\infty} a_n$ converges if*

$$\lim n\left(\frac{a_n}{a_{n+1}} - 1\right) > 1$$

and diverges if

$$\lim n\left(\frac{a_n}{a_{n+1}} - 1\right) < 1.$$

Proof In the first case, there is an $r > 0$ such that for n sufficiently large,

$$n\left(\frac{a_n}{a_{n+1}} - 1\right) > 1 + r.$$

Take $c_n = n$ in Kummer's Test.

In the second case, for n sufficiently large, we have

$$n\left(\frac{a_n}{a_{n+1}} - 1\right) < 1$$

or

$$\frac{a_n}{a_{n+1}}n - (n+1) < 0.$$

Again, take $c_n = n$ in Kummer's Test. □

Raabe's Test frequently settles the case where the limiting ratio is

$$\lim \frac{a_{n+1}}{a_n} = 1$$

and is therefore stronger than the Ratio Test.

Example 1 For the series

$$\sum_{n=1}^{\infty} \frac{1}{n^p}$$

we have

$$\lim n\left(\frac{a_n}{a_{n+1}} - 1\right) = \lim n\left(\frac{(n+1)^p}{n^p} - 1\right)$$
$$= \lim n\left(\left(1 + \frac{1}{n}\right)^p - 1\right) = p.$$

Thus by Raabe's Test, the series converges if $p > 1$ and diverges if $p < 1$. However, Raabe's Test does not decide the borderline case $p = 1$. ∎

A consequence of Kummer's Test that *does* decide the case $p = 1$ is *Gauss's Test*. For the statement, we will use a notation due to G. H. Hardy. Given two sequences a_n and b_n, we say that

$$b_n = O(a_n)$$

iff

$$|b_n| \leq C a_n$$

for some constant C.

Theorem 10 (Gauss's Test) *Let $a_n > 0$. Suppose that there exists a positive number p such that*

$$\frac{a_n}{a_{n+1}} = 1 + \frac{p}{n} + O(\frac{1}{n^2}).$$

Then $\sum_{n=1}^{\infty} a_n$ converges if $p > 1$ and diverges if $p \leq 1$.

Proof If $p \neq 1$, take $c_n = n$ in Kummer's Test. If $p = 1$, take $c_n = n \log n$. □

Example 2 For the series

$$\sum_{n=1}^{\infty} \frac{1}{n^p}$$

we have

$$\frac{a_n}{a_{n+1}} = \frac{(n+1)^p}{n^p} = \left(1 + \frac{1}{n}\right)^p = 1 + \frac{p}{n} + O\left(\frac{1}{n^2}\right).$$

Thus by Gauss's Test, the series converges if $p > 1$ and diverges if $p \leq 1$. ∎

9.4.1 Problems

1. Show by Gauss's Test that

$$\sum_{n=1}^{\infty} \frac{1 \cdot 3 \cdot 5 \cdots (2n-1)}{2 \cdot 4 \cdot 6 \cdots (2n)} = 1 + \frac{1 \cdot 3}{2 \cdot 4} + \frac{1 \cdot 3 \cdot 5}{2 \cdot 4 \cdot 6} + \cdots$$

is divergent, while

$$\sum_{n=1}^{\infty} \frac{1 \cdot 3 \cdot 5 \cdots (2n-1)}{2 \cdot 4 \cdot 6 \cdots (2n)} \cdot \frac{1}{2n+1} = 1 + \frac{1 \cdot 3}{2 \cdot 4} \cdot \frac{1}{5} + \frac{1 \cdot 3 \cdot 5}{2 \cdot 4 \cdot 6} \cdot \frac{1}{7} + \cdots$$

is convergent.

2. Use Stirling's approximation

$$n! \sim \sqrt{2\pi n} \; n^n e^{-n}$$

to solve Problem 1.

Appendix A

Logic, Sets, and Functions

A.1 Logic

First, we note that, as always in Mathematics, the statement "*P or Q*" means "*P or Q, or both.*" In those cases where we mean "*P or Q, but not both*," we will say "*P or Q, but not both.*"

The *converse* of a statement of the form "*If P, then Q*" is the statement "*If Q, then P.*" These statements are by no means the same; one can be true but not the other. For example, the converse of

$$\text{"If } x = 2, \text{ then } x^2 = 4\text{"}$$

is

$$\text{"If } x^2 = 4, \text{ then } x = 2.\text{"}$$

The first statement is true; the second is not.

To say that "*P if, and only if Q*" is equivalent to asserting that both "*If P, then Q*" and "*If Q, then P*" are true. Following the now common usage of Paul Halmos, we shall write "*P iff Q*" for "*P if, and only if Q.*"

A.2 Sets

According to G. Cantor, a *set* is "*a collection into a whole of definite, well defined objects of our thought or perception.*" These objects are called the *elements* of the set. We write $a \in A$ to denote that a is an element of A.

A set is determined by its elements; that is, two sets A and B are *equal* iff they have the same elements. To show that two sets are equal is therefore to show that every element of A is also an element of B, and then to show that every element of B is also an element of A.

A set A is a *subset* of a set B—in symbols $A \subset B$—iff every element of A is also an element of B. Thus, $A = B$ iff $A \subset B$ and $B \subset A$.

Given two sets A and B, their *union* $A \cup B$ is the set of all elements that are either in A or in B or in both; their *intersection* $A \cap B$ is the set of all elements that are in both sets; and their *difference* is the set $A \setminus B$ of all elements of A that are not in B. If A is understood to be a subset of some larger "universal" set Ω, such as, for example, the Real Numbers, the complement of A (relative to Ω) is the set $A^c = \Omega \setminus A$.

We may consider as well the union and intersection of infinite collections of sets:

$$\bigcup_{\alpha \in I} A_\alpha \text{ and } \bigcap_{\alpha \in I} A_\alpha$$

where I is an index set. Unions and intersections satisfy *De Morgan's Laws*:

$$\left(\bigcup_{\alpha \in I} A_\alpha\right)^c = \bigcap_{\alpha \in I} A_\alpha^c \text{ and}$$

$$\left(\bigcap_{\alpha \in I} A_\alpha\right)^c = \bigcup_{\alpha \in I} A_\alpha^c.$$

If $P(x)$ is a statement about x, we use the notation $\{x : P(x)\}$ to denote the set of all elements such that $P(x)$ is true. For example, the set

$$\{x : x \text{ is a real number and } 0 < x < 1\}$$

is the open interval $(0, 1)$.

In this notation, we can write

$$A \cup B = \{x : x \in A \text{ or } x \in B\}$$
$$A \cap B = \{x : x \in A \text{ and } x \in B\}$$
$$A \setminus B = \{x : x \in A \text{ and } x \notin B\}$$

and similarly,

$$\bigcup_{\alpha \in I} A_\alpha = \{x : x \in A_\alpha \text{ for some } \alpha \in I\}$$
$$\bigcap_{\alpha \in I} A_\alpha = \{x : x \in A_\alpha \text{ for all } \alpha \in I\}$$

A.3 Functions

A *function* f from a set A into a set B is a rule that assigns to each element a of A a unique element $f(a)$ in B. We call $f(a)$ the *value* of f at a.

A function may be thought of as a *mapping* (or just a *map*) from A into B, in which case we refer to $f(a)$ as the *image of a under f*. One sometimes writes $f: A \longrightarrow B$ to mean that f is a function from A into B.

The set A of all elements a for which the value $f(a)$ is defined is called the *domain* of f. The *range* $ran(f)$ of f is the set of all elements of B that are images of some point of A under f.

$$ran(f) = \{b \in B : b = f(a) \text{ for some } a \in A\}.$$

More generally, if S is a subset of A, the *image of S under f* is the set $f(S)$ of all points of B that are images of points of S under f; in symbols,

$$f(S) = \{b \in B : b = f(a) \text{ for some } a \in S\}.$$

In particular, the range of f is $ran(f) = f(A)$.

If S is a subset of the range space B, the *inverse image of S under f* is the set of all elements of A that are mapped by f into a point of the set S; in symbols:

$$f^{-1}(S) = \{a \in A : f(a) \in S\}.$$

The map $f^{-1}(A)$ preserves unions and intersections:

$$f^{-1}(A \cup B) = f^{-1}(A) \cup f^{-1}(B)$$
$$f^{-1}(A \cap B) = f^{-1}(A) \cap f^{-1}(B).$$

We say that the function f maps A *onto* B iff every element of B is the image of an element of A; in symbols, $f(A) = B$.

The function f is *one-to-one* iff no two elements of A map to the same element of B; that is, if $f(a) = f(a')$ implies that $a = a'$.

A one-to-one function has an *inverse function* $f^{-1}(b)$ defined on its range $f(A)$ such that

$$f(f^{-1}(b)) = b \text{ for every } b \in f(A)$$
$$f^{-1}(f(a)) = a \text{ for every } a \in A.$$

Do not confuse the inverse function $f^{-1}(b)$, which exists only when f is one-to-one, with $f^{-1}(S)$ for a subset S of B, which always exists.

A.3.1 Problems

1. What are the converses of these statements?

   (a) If $x^2 = 4$, then x is rational.

   (b) If $f(x)$ is differentiable, then $f(x)$ is continuous.

2. Prove DeMorgan's laws.

3. Let $f : A \longrightarrow B$.

   (a) Prove that if $S \subset A$ then
   $$S \subset f^{-1}(f(S)).$$

   (b) Prove that if f is one-to-one, then
   $$S = f^{-1}(f(S)).$$

   (c) Show by an example that the two sets are not equal in general.

4. Are either of these statements true?

   (a) $f(A \cup B) = f(A) \cup f(B)$
   (b) $f(A \cap B) = f(A) \cap f(B)$.

A.4 Countable and Uncountable Sets

Consider the sets $\mathbb{Z}$ of integers, $\mathbb{Q}$ of rationals, and set $\mathbb{R}$ of reals. They are all infinite sets, but are they the same size? In order to answer such a question, we must first ask what it means to say that two sets are the same size.

These questions were asked by G. Cantor. His answer was that two sets are the same size if their elements could be put into one-to-one correspondence.

Definition 1 Two sets A and B are *equivalent* iff there is a one-to-one function f from A onto B.

A one-to-one mapping of A onto B is called a *bijection*.

Thus, for example, the sets of odd and even numbers are equivalent, since the function
$$f(n) = n + 1$$

is a bijection from the odds to the evens. We can display the correspondence as follows:

$$1, 3, 5, 7, \ldots$$
$$2, 4, 6, 8, \ldots.$$

More surprisingly, although the set of even numbers is a subset of the set of natural numbers, the two sets are equivalent. In fact, the function

$$f(n) = 2n$$

is a bijection from the natural numbers to the evens. We can display this as

$$1, 2, 3, 4, \ldots$$
$$2, 4, 6, 8, \ldots.$$

Indeed, one definition of an infinite set is that it is equivalent to a subset of itself.

Definition 2 A set A is *countable* iff it is either finite or equivalent to the set $\mathbb{N}$ of natural numbers. A set which is not countable is said to be *uncountable*.

Thus we have proved

Theorem 1 *The sets of even and odd numbers are countable.*

The question naturally arises as to whether there are any sets that are not countable. One might think since there are an infinite number of rationals between every pair of integers, that $\mathbb{Q}$ is a much larger set than $\mathbb{N}$ and might be uncountable, but this is not so.

Theorem 2 *The set $\mathbb{Q}$ of rationals is countable.*

Proof We will write the positive rationals in sequential order. To do this, first write the positive rationals with denominator 1 in increasing order on a line. Underneath, write those with denominator 2 in the same way. Continue with the other denominators. Then arrange the totality of rationals in sequential order by following the arrows in the diagram below, omitting any numbers that have already appeared in the list.

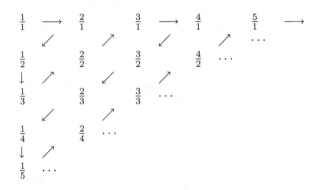

This method is called *Cantor's first diagonal argument*. □

Theorem 3 *The interval $(0,1)$ is not countable.*

Proof Suppose on the contrary, that the real numbers on the interval $(0,1)$ are countable. Then they can be listed in sequence as follows.

$$a_1 = 0.a_{11}a_{12}a_{13}\cdots a_{1n}\cdots$$
$$a_2 = 0.a_{21}a_{22}a_{23}\cdots a_{2n}\cdots$$
$$a_3 = 0.a_{31}a_{32}a_{33}\cdots a_{3n}\cdots$$
$$\vdots$$
$$a_n = 0.a_{n1}a_{n2}a_{n3}\cdots a_{nn}\cdots$$
$$\vdots$$

The right side denotes the unique non-terminating decimal expansion of each number.

Define a real number

$$x = 0.x_1x_2x_3\cdots x_n\cdots$$

with the decimal expansion

$$x_n = 5 \quad \text{if } a_{nn} \neq 5$$
$$x_n = 4 \quad \text{if } a_{nn} = 5.$$

Then x cannot appear in the list, since it differs from a_n in the nth decimal place. Hence, the real numbers cannot be listed, and are uncountable.

This method is known as *Cantor's second diagonal argument*. □

Theorem 4 *The reals are equivalent to $(0, 1)$, and are hence uncountable.*

Proof A bijection from $\mathbb{R}$ onto $(0, 1)$ is given by

$$f(x) = \frac{2}{\pi} \arctan(x).$$ □

We will stop at this point. However, there is an extensive theory of infinite—or *transfinite*—numbers. There turns out to be a whole hierarchy of transfinite numbers, and the reals are by no means the largest. (In fact, there is no largest.)

A.4.1 Problems

1. Prove that the set of integers $\mathbb{Z}$ is countable.

2. Write $A \approx B$ iff A and B are equivalent. Prove that

    (a) $A \approx A$.
    (b) If $A \approx B$, then $B \approx A$.
    (c) If $A \approx B$, and $B \approx C$, then $A \approx C$.

Appendix B
The Topology of $\mathbb{R}$

B.1 Introduction

This appendix is concerned with the General Topology of subsets of the real line. We will introduce the notions of *open* and *closed* sets, and will characterize *continuous functions* in terms of open sets.

We will then introduce the important notion of a compact set, and will prove the *Heine–Borel Theorem*, which tells exactly which subsets of $\mathbb{R}$ are compact. We will then use the Heine–Borel Theorem to discuss the *Extreme Value Theorem* and uniform continuity.

The reader will no doubt find this subject considerably more abstract than what has preceded it, and may quite reasonably ask why we are doing this. The answer is that we are dealing here with a simple case of a very general theory. The notions of open sets, continuity, and compactness can be applied to many systems other than the real numbers, and provide a means of thinking about the common properties of very different things in a unified way.

B.2 Open Sets

The basic concept of general topology is that of an *open set*.

Definition 1 A *neighborhood of a point* a is an open interval containing a.

If $r > 0$, we write
$$B(a, r) = \{x : |x - a| < r\}.$$

$B(a, r)$ is just the interval $(a - r, a + r)$. $B(a, r)$ is then a neighborhood of a, and a is its center.

Definition 2 A point a is an *interior point* of a set S iff S contains a neighborhood of a.

Definition 3 A set U is said to be *open* iff every point of U is an interior point of U; that is, iff U contains a neighborhood of every point of U.

By definition, the empty set $\emptyset$ will be considered an open set.

Example 1 Any open interval (a,b) is an open set. The whole line $\mathbb{R}$ and any open half-line (a,∞) or $(-\infty,a)$ is an open set. On the other hand, the half-open interval $[a,b)$ is not open, because it contains no neighborhood of the point a. ∎

Theorem 1 *The union of an arbitrary collection of open sets is open.*

Proof Let
$$U = \bigcup_\alpha U_\alpha$$
where the U_α are open. If $p \in U$, then $p \in U_\alpha$ for some α. Choose a neighborhood $B(p,r) \subset U_\alpha$. Then $B(p,r) \subset U$. □

Example It follows that any finite or infinite union of open intervals is an open set. For example, the sets
$$(0,1) \cup (2,3)$$
and
$$(0,1) \cup (2,3) \cup (4,5) \cup \ldots$$
are open. In fact, it can be shown that a set U is an open set in $\mathbb{R}$ iff U is the union of a sequence of disjoint open intervals. ∎

Theorem 2 *The intersection of a finite number of open sets is open.*

Proof Let
$$U = U_1 \cap \ldots \cap U_n$$
where $U_1, \ldots, U_n$ are open. If $p \in U$, then $p \in U_k$ for every k. Choose a neighborhood $B(p, r_k) \subset U_k$. If $r = \min(r_1, \ldots, r_n)$, then $B(p,r) = B(p, r_1) \cap \ldots \cap B(p, r_n) \subset U$. □

The intersection of an infinite number of open sets need not be open. For example, the intersection of the sequence $U_n = (-1/n, 1/n)$ of open sets is just $\{0\}$, which is not open.

B.2.1 Continuity and Open Sets

Continuous functions can be described in terms of open sets.

Definition 4 Let S be a subset of $\mathbb{R}$. A subset A of S is said to be *open in S* (or *relatively open* in S) iff there is an open set U in $\mathbb{R}$ such that $A = S \cap U$.

Theorem 3 *A function $f(x)$ is continuous on S if $f^{-1}(U)$ is open in S for every open set U of $\mathbb{R}$.*

Proof Suppose that $f(x)$ is continuous. Let U be open, and a a point of $f^{-1}(U)$. Since U is open, there is a neighborhood $B(f(a), \epsilon)$ of $f(a)$ contained in U. Since f is continuous, there exists $\delta > 0$ such that $|x - a| < \delta$ implies that $|f(x) - f(a)| < \epsilon$, or in other words, that $x \in B(a, \delta)$ implies that $f(x) \in B(f(a), \epsilon)$. Therefore, $B(a, \delta) \subset f^{-1}(U)$.

Conversely, suppose that the inverse image under f of an open set is always open. Let $a \in S$, and $\epsilon > 0$. Then $f^{-1}(B(f(a), \epsilon))$ is open in S, and so must contain a neighborhood $B(a, \delta) \cap S$ of a in S. This means that $|x - a| < \delta$ (and $x \in S$) implies that $|f(x) - f(a)| < \epsilon$. □

Careful inspection of this proof will show that the two statements are just two ways of saying the same thing. You may wonder why one bothers with such things. The answer is that the second condition is more readily generalized to other situations.

B.2.2 Problems

1. Prove that an open interval is an open set.

2. Which of the following sets are open?

    (a) $\{x : x^2 > 1\}$
    (b) $\{x : x^2 + x + 1 \geq 2\}$
    (c) $\{x : \sin x > 0\}$
    (d) $\{2, 3\}$
    (e) $\{x : x^2 + 1 \leq 0\}$

3. Show that the intersection of an infinite collection of open sets need not be open. (*Hint*: Consider $U_n = (-1/n, 1/n)$).

*4. The *interior of a set* A is the set of all interior points of A:
$$int\,(A) = \{x \in A : x \text{ is an interior point of } A\}.$$

Prove that $int\,(A)$ is an open set, and that it is the largest open set contained in A.

5. Find the interior of these sets.

   (a) $\{x : x^2 > 1\}$
   (b) $\{x : x^4 + 2x^2 \geq 3x^3\}$
   (c) $\{x : \sin x \geq 0\}$
   (d) $\{2, 3\}$
   (e) The set $\mathbb{Q}$ of rationals
   (f) $\{x : x^2 + 1 \leq 0\}$

B.3 Closed Sets

The complement of an open set is called a closed set.

Definition 5 *A set K is closed iff its complement K^c is open.*

Theorem 4 *The intersection of an arbitrary collection of closed sets is closed.*

Proof Let
$$K = \bigcap_\alpha K_\alpha$$
where the K_α are closed. By DeMorgan's Law
$$K^c = \bigcup_\alpha K_\alpha^c.$$
But K_α^c is open because K_α is closed, so that K^c is open by Theorem 1. □

Theorem 5 *The union of a finite number of closed sets is closed.*

Proof Let
$$K = K_1 \cup \ldots \cup K_n$$
where $K_1, \ldots, K_n$ are closed. By DeMorgan's Law
$$K^c = K_1^c \cap \ldots \cap K_n^c.$$
But $K_1^c, \ldots, K_n^c$ are open, so that K^c is open by Theorem 2. □

Examples

(a) The empty set $\emptyset$ and the whole line $\mathbb{R}$ are closed. It can be shown that these are the only two sets in $\mathbb{R}$ that are both open and closed.

(b) Any finite set is closed; in particular, a set $\{p\}$ consisting of a single point p is closed.

(c) A closed interval
$$[a,b] = \{x : a \leq x \leq b\}$$
is closed.

(d) The half-open intervals $(a,b]$ and $[a,b)$ are *neither open nor closed*.

(e) The set
$$\left\{1, \frac{1}{2}, \frac{1}{3}, \frac{1}{4}, \ldots\right\} \cup \{0\}$$
consisting of the sequence $1, \frac{1}{2}, \frac{1}{3}, \frac{1}{4}, \ldots$ and its limit is closed. ∎

B.3.1 The Closure of a Set

Definition 6 *The* closure *$\bar{S}$ of a set S is the intersection of all closed sets containing S.*

Theorem 6 *The closure of a set S is closed, and is the smallest closed set containing S.*

Proof $\bar{S}$ clearly contains S, since it is the intersection of a family of sets all of which contain S. Moreover, $\bar{S}$ is closed because it is the intersection of a family of closed sets. If K is any closed set containing S, then $\bar{S} \subset K$ because K is one of this family of sets. Thus, $\bar{S}$ is the smallest such set. □

Corollary 1 *A set K is closed iff $K = \bar{K}$.*

Examples

(a) Any closed set is its own closure. Thus, empty set $\emptyset$ is the closure of $\emptyset$, $\{p\}$ is the closure of $\{p\}$, and $[a,b]$ is the closure of $[a,b]$.

(b) The closed interval $[a,b]$ is also the closure of (a,b), $[a,b)$, and $(a,b]$.

(c) The closure of the set
$$S = \left\{1, \frac{1}{2}, \frac{1}{3}, \frac{1}{4}, \ldots\right\}$$
consisting of the sequence $1, \frac{1}{2}, \frac{1}{3}, \frac{1}{4}, \ldots$ is obtained by adding the limit 0 to the set
$$\bar{S} = \left\{1, \frac{1}{2}, \frac{1}{3}, \frac{1}{4}, \ldots\right\} \cup \{0\}.$$
∎

Theorem 7 *A point p is in the closure of a set K iff every neighborhood of p contains a point of K.*

Proof If $p \notin \bar{S}$, then $p \in (\bar{S})^c$. Since $\bar{S}$ is closed, $(\bar{S})^c$ is open, and so there is a neighborhood N of p contained in $(\bar{S})^c$. But then N can contain no point of S.

Conversely, suppose there is a neighborhood N of p that contains no point of S. Since N is open, N^c is closed, so that $\bar{S} \setminus N = \bar{S} \cap N^c$ is a closed set containing S. By definition, $\bar{S} \subset \bar{S} \cap N^c$, which means that N does not intersect $\bar{S}$; so in particular, $p \notin \bar{S}$. □

Theorem 8 *A point p is in the closure $\bar{S}$ of a set S iff there exists a sequence $x_n \in S$ such that $x_n \to p$.*

Proof If $p \in \bar{S}$, then by Theorem 7, $N(p, 1/n)$ contains a point p_n. But then $|p_n - p| < 1/n$, which implies that $p_n \to p$. □

Corollary 2 *A set K is closed iff $x_n \in K$ and $x_n \to p$ implies $p \in K$.*

Thus, the closure of a set S is obtained from S by adding to S all points that are limits of sequences in S; or, in other words, which can be approximated arbitrarily closely by points of S.

B.3.2 Problems

1. Prove that the closed interval $[a, b]$ is closed.

2. Prove that the half-open interval $(a, b]$ is neither open nor closed.

3. Which of the following sets are closed?

    (a) $\{x : x^2 \geq 1\}$

(b) $\{x : x^2 + x + 1 = 2\}$

(c) $\{x : \sin x > 0\}$

(d) $\{2, 3\}$

(e) $[0, 1)$

(f) $\{1/2, 2/3, \cdots, n/(n-1), \cdots\}$

(g) $\{x : \sin(1/x) = 0\}$

4. Show that the union of an infinite collection of closed sets need not be closed. (*Hint:* Consider $K_n = \left[\frac{1}{n}, 1 - \frac{1}{n}\right]$.)

*5. Prove that the empty $\emptyset$ set and the whole line $\mathbb{R}$ are the only two subsets of $\mathbb{R}$ that are both open and closed.

6. Find the closures of the following sets.

   (a) $\{x : x^2 > 1\}$

   (b) $\{x : x^2 + x + 1 = 2\}$

   (c) $\{x : \sin(1/x) > 0\}$

   (d) $\{x : \sin(1/x) = 0\}$

   (e) $\{x : 0 < x < 1 \text{ and } x \text{ rational}\}$

   (f) $[0, 2)$

7. Prove that

   (a) The supremum $\sup S$ of a set is an element of its closure $\bar{S}$.

   (b) Prove that a closed, bounded set has a largest element.

8. Prove Corollary 1.

9. Prove Corollary 2.

10. Prove that
$$\bar{S} = [int(S^c)]^c.$$

Dense Sets   A set D is *dense* in a set S iff $S \subset \bar{D}$.

11. Prove that the set $\mathbb{Q}$ of rationals is dense in the set of reals $\mathbb{R}$.

12. Let the two functions $f(x)$ and $g(x)$ be defined and continuous on a set S, and let D be dense in S. Prove that if
$$f(x) = g(x)$$

for every $x \in D$, then $f(x)$ and $g(x)$ are equal at all points of S.

Accumulation Points A point p is an *accumulation point* of a set S iff every neighborhood of p contains a point of S different from p. An accumulation point of S need not be in the set S itself.

A point p is an *isolated point* of a set S iff $p \in S$ but p is not an accumulation point of S.

13. What are the accumulation points and isolated points of the following sets?

    (a) $\{x : x^4 + 2x^2 \geq 3x^3\}$
    (b) $\{1, \frac{1}{2}, \frac{1}{3}, \frac{1}{4}, \dots\}$
    (c) $\{2, 3\}$
    (d) $[0, 1)$
    (e) The set of rationals $\mathbb{Q}$

14. Prove that a set is closed iff it contains all of its accumulation points.

    Boundary Points A point p is a *boundary point* of a set S iff every neighborhood of p contains a point of S and a point of its complement S^c.

    The *boundary of a set* is the set of all its boundary points. The boundary of a set is denoted by ∂S.

15. What are the boundaries of the following sets?

    (a) $\{x : x^4 + 2x^2 \geq 3x^3\}$
    (b) $\{1, \frac{1}{2}, \frac{1}{3}, \frac{1}{4}, \dots\}$
    (c) $\{2, 3\}$
    (d) $[0, 1)$
    (e) The set $\mathbb{Q}$ of rationals
    (f) The $\mathbb{R}$ set of reals
    (g) The empty set

16. Prove that
$$\partial S = \overline{S} \cap \overline{S^c}.$$

17. Prove that
$$\partial S = \partial S^c.$$

B.4 Compact Sets

The notion of a *compact set* is extremely important in modern analysis.

Definition 7 An *open cover* of a set S is a collection $\{U_\alpha\}$ of open sets such that S is contained in their union:

$$S \subset \bigcup_\alpha U_\alpha.$$

A *finite subcover* of $\{U_\alpha\}$ is a finite collection $U_{\alpha_1}, U_{\alpha_2}, \cdots, U_{\alpha_n}$ of sets of $\{U_\alpha\}$ that also covers S; that is

$$S \subset U_{\alpha_1} \cup U_{\alpha_2} \cup \cdots \cup U_{\alpha_n}.$$

Definition 8 A set K is *compact* iff every open cover contains a finite subcover of S.

Examples The empty set and any finite set are compact. The open interval $(0,1)$ is not compact, since the open cover

$$U_n = (1/n, 1) \qquad n = 1, 2, 3, \ldots$$

has no finite subcover. We will shortly see that the unit interval $[0,1]$ is compact. ∎

Theorem 9 *A closed subset of a compact set is compact.*

Proof Let C be a closed subset of the compact set K. Let $\{U_\alpha\}$ be an open cover of C. Since C is closed, its complement C^c is open. Adding C^c to $\{U_\alpha\}$ therefore gives an open cover

$$\{U_\alpha\} \cup \{C^c\}$$

of K. But K is compact, so there is a finite subcover $U_1, \ldots, U_n, C^c$ of K such that

$$C \subset K \subset (U_1 \cup \ldots \cup U_n \cup C^c).$$

But then

$$C \subset (U_1 \cup \ldots \cup U_n)$$

since C and C^c are disjoint. □

B.4.1 The Heine–Borel Theorem

The Heine–Borel Theorem is a very important and powerful result, comparable to the Bolzano–Weierstrass Theorem.

Theorem 10 *A compact subset of $\mathbb{R}$ is closed and bounded.*

Proof We will show that the complement of K is open. Let p be a point in K^c; that is, $p \notin K$. For each point $x \in K$, let U_x be a neighborhood of x such that $p \notin U_x$, and V_x a neighborhood of p with $U_x \cap V_x = \emptyset$. The collection $\{U_x : x \in K\}$ is an open cover of K, since any point x in K is in U_x. It must therefore contain a finite subcover $U_{x_1}, \ldots, U_{x_n}$ of K. Now $V = V_{x_1} \cap \ldots \cap V_{x_n}$ is a neighborhood of p contained in K^c. For V does not intersect K since

$$K \subset U_{x_1} \cup \ldots \cup U_{x_n}.$$

Hence, K^c is open, and so K is closed.

To see that K is bounded, note that the collection of open intervals

$$\{I_n = (-n, n) : n = 1, 2, \ldots\}$$

is an open cover of K. By compactness there is a finite subcover $I_1, \ldots, I_N$. But since I_n is an increasing sequence of intervals, we have $K \subset I_N$ for some N. Thus $|x| \leq N$ for all $x \in K$, so that K is bounded. $\square$

Theorem 11 *A closed finite interval $[a, b]$ is a compact set.*

Proof Assume for simplicity that $[a, b]$ is the unit interval $[0, 1]$. Let $\{U_\alpha\}$ be an open cover of $[0, 1]$. Let

$$S = \{t : 0 \leq t \leq 1 \text{ and } \{U_\alpha\} \text{ has a finite subcover of } [0, t]\}$$

and $c = \sup S$. Clearly, $0 \leq c \leq 1$, so c is in some member U_0 of $\{U_\alpha\}$. Since U_0 is open, it contains an interval about c, and hence two points s and t such that $t < c < s$. Now since $t < c$, $[0, t]$ has a finite subcover $U_1, \ldots, U_n$. But then $[0, s]$ also has the finite subcover $U_0, U_1, \ldots, U_n$. This implies that $s \in S$ and $s > c$, a contradiction, unless $c = 1$, in which case, s is not in S. $\square$

The Heine–Borel Theorem tells exactly which subsets of $\mathbb{R}$ are compact.

Theorem 12 (Heine–Borel Theorem) *A subset of the real numbers is compact iff it is closed and bounded.*

Proof By Theorem 9, we need only show that if K is closed and bounded, then K is compact. Since K is bounded, it is a subset of some closed, finite interval $[a, b]$. Therefore, K is a closed subset of $[a, b]$. But $[a, b]$ is compact by Theorem 11, and therefore K is compact by Theorem 8. $\square$

B.4.2 Compactness and Continuity

We will illustrate the use of the Heine–Borel Theorem by using it to prove two theorems previously proved using the Bolzano–Weierstrass Theorem. The first is the Extreme Value Theorem.

Theorem 13 *The continuous image of a compact set is compact.*

Proof Let f be continuous on the compact set K. Let $\{U_\alpha\}$ be an open cover of $f(K)$. Then, since f is continuous, $\{f^{-1}(U_\alpha)\}$ is an open cover of K. So there must exist a finite subcover

$$f^{-1}(U_{\alpha_1}), f^{-1}(U_{\alpha_2}), \ldots, f^{-1}(U_{\alpha_n})$$

of K. But then $U_{\alpha_1}, U_{\alpha_2}, \ldots, U_{\alpha_n}$ cover $f(K)$. $\square$

Theorem 14 (Extreme Value Theorem) *A continuous function on a compact set K attains its maximum value at some point of K.*

Proof If K is compact, then $f(K)$ is also compact by Theorem 13. By Theorem 12, K is bounded, so $M = \sup f(K)$ is finite. But $f(K)$ is also closed, again by Theorem 12, and so contains its supremum M. Thus $M = f(a)$ for some $a \in K$. $\square$

Note that the theorem in this form is more general than the previous version, since it applies to an arbitrary compact set, rather than just a closed interval. The same applies to the next result—the theorem on *uniform continuity*.

Theorem 15 *A continuous function on a compact set K is uniformly continuous.*

Proof Let $\epsilon > 0$. Since f is continuous, for each a in K there is a $\delta(a)$ such that $|f(x) - f(a)| < \epsilon/2$ when $|x - a| < \delta(a)$. The neighborhoods

$$\left\{ B\left(a, \frac{\delta(a)}{2}\right) : a \in K \right\}$$

cover K, and so by compactness, a finite number of them,

$$B\left(a_1, \frac{\delta(a_1)}{2}\right), \ldots, B\left(a_n, \frac{\delta(a_n)}{2}\right)$$

cover K. Take

$$\delta = \frac{1}{2} \min \{\delta(a_1), \ldots, \delta(a_n)\}.$$

Let $|x-y| < \delta$, and suppose that $x \in B(a_k, \delta(a_k)/2)$. Then

$$|y - a_k| \leq |y - x| + |x - a_k| < \frac{\delta(a_k)}{2} + \delta < \delta(a_k)$$

so that x and y are both in $B(a_k, \delta(a_k))$. Hence,

$$|f(x) - f(y)| \leq |f(x) - f(a_k)| + |f(y) - f(a_k)| < \frac{\epsilon}{2} + \frac{\epsilon}{2} = \epsilon$$

when $|x - y| < \delta$. □

B.4.3 Problems

1. Prove directly from the definition that a finite set is compact.

2. Show directly by example that the whole line $\mathbb{R}$ is not compact.

3. Completeness is necessary for Theorem 11. Where is it used in its proof?

4. Show that Theorem 11 fails for the Rationals by giving an open cover of the set

$$S = \mathbb{Q} \cap [0, 1]$$

with no finite subcover.

Appendix C
Recommended Reading

C.1 Introduction

Calculus

The following are three excellent rigorous introductions to Calculus. All three are good sources for challenging problems.

Hardy, G. H. (1993). *A course of pure mathematics* (10th ed.). Cambridge, UK: Cambridge University Press.

A classic by one of the great mathematicians of the 20th century. The first edition appeared in 1908. It features problems from the famous—or perhaps notorious—Cambridge Tripos exams.

Courant, R. (1966). *Differential and integral calculus* (Vol. 1, 2). Wiley reprint. Hoboken, NJ.

A translation of Courant's lectures from the early 1930s by my former colleague, the late E. J. McShane. Courant's style is somewhat unique, but once one gets used to it, the book is unexcelled. The problems, due to McShane, are frequently difficult. As was customary, the German edition lacked problems. Courant's instructions to McShane were to add problems, but none that were "trivial."

Apostol, T. M. (1967, 1969). *Calculus* (Vol. I, II). Hoboken, NJ: Wiley.

The high point of the late 1950s' and early 1960s' movement toward rigor in introductory Calculus, from which we have been in retreat ever since. The second edition of Volume I had, to my taste, too much real variable for a beginning student, and as an actual introduction to Calculus, I preferred the first edition. As a supplement to the present text, however, the second edition may be more useful.

Foundations of the Number System

Landau, E. (1966). *Foundations of analysis.* Providence, RI: American Math. Soc. 2001.

A classic. Starting from Peano's Axioms for the natural numbers, it constructs the real number system by defining a real number to be a Dedekind cut of the rationals. Very concise, consisting entirely of definitions, theorems, and proofs.

Dedekind, R. (1963). *Essays on the theory of numbers.* Dover reprint. Mineola, NY.

Translations of two of Dedekind's original memoirs, one on the natural numbers, and the one in which cuts are introduced. Beautifully clear, and not to be missed.

Introductory Real Analysis

The next two books treat the present subject at a somewhat more sophisticated level. They also discuss functions of several variables, and contain an introduction to Lebesgue theory. They are very different. As one reviewer put it, Rudin is brief and elegant; Apostol is thorough and detailed. Each is excellent in its own way.

Apostol, T. M. (1974). *Mathematical analysis* (2nd ed.). Reading, MA: Addison-Wesley.

Rudin, W. (1976). *Principles of mathematical analysis* (3rd ed.). New York: McGraw-Hill.

Boas, R. (1997). Primer of Real Functions. *Carus mathematical monographs* (No. 13, 4th ed.). Mathematical Association of America. Washington, D.C.

A classic discussion of selected topics in real analysis.

Infinite Series

Hirschman, I. I. (1978). *Infinite series.* Greenwood Press reprint. Santa Barbara, CA.

A very nice, concise introduction to the subject.

Knopp, K. (1956). *Infinite series.* Dover reprint. Mineola, NY.

A classic work on infinite series.

Set Theory and Transfinite Numbers

Halmos, P. R. (1998). *Naive set theory*. Springer: Undergraduate texts in Mathematics. New York.

An elementary intuitive introduction to set theory.

Suppes, P. (1972). *Axiomatic set theory*. Dover reprint.

A more advanced treatment that is more careful about axioms.

Index

A
Abel's summation by parts formula, 179
Abel's Theorem, 181
absolute convergence, 42, 113, 137
absolute value, 7
accumulation point, 206
alternating series, 140
Archimedean Property, 15, 16, 24, 39
arithmetic mean, 117
associative law, 5
Axiom
    Algebra, 5
    Continuity, 8
    Order, 6

B
Bernoulli's inequality, 4, 33
Bernstein's Theorem, 185
beta function, 118
bijection, 194
Binomial Theorem, 4
    Newton's, 186
bisection method, 37, 62
Bolzano-Weierstrass Theorem, 37, 39, 40, 65, 66, 100, 161, 208
boundary of a set, 206
boundary point, 206
bounded function, 65
bounded sequence, 27
Brouwer Fixed Point Theorem, 68

C
Cantor's diagonal arguments, 196
Cantor's Principle, 32, 39, 40, 63
Cauchy's Convergence Criterion
    for functions, 67
    for sequences, 40, 42,
Cauchy sequence, 39–40, 42, 102, 138
Cauchy's Mean Value Theorem, 79–81, 83, 108
Cesaro limit, 45, 182
chain rule, 73, 109
closed set, 202
closure of a set, 203
commutative law, 5
compact set, 207
comparison test, 130, 133, 153
complete induction, 18
composition
    differentiation of, 73
    of continuous functions, 56
    of integrable functions, 98
conditional convergence, 114, 139
continuity
    and open sets, 201
    sequential criterion for, 55
convergence
    interval of, 164
    necessary condition for, 127
    of improper integral, 112
    of sequence, 24
    of series, 123
    pointwise, 149
    radius of, 164
    uniform, 149, 153, 163
converse of a statement, 191
countable set, 195
Critical Point Theorem, 75, 77

D

Darboux sum, 88
Dedekind cut, 8, 10, 16, 212
Dedekind's Axiom, 8, 11, 17, 34, 39
De Morgan's laws, 192, 202
dense set, 15, 205
derivative, 69
diagonal arguments, 196
difference of sets, 192
Dini's Theorem, 161
Dirichlet's Test, 180
distributive law, 5
domain, 193
double limit, 159

E

e, Euler's number, 33
    computation of, 175
    irrationality of, 176
elliptic integral, 119
equivalent sets, 194
estimation, 25
Euler–Mashceroni constant, 120
exponential series, 164, 171
Extreme Value Theorem, 64–66, 77, 209

F

Fibonacci numbers, 22, 47
field, 5, 19
    ordered field, 7, 19
Fresnel integral, 115
Frulanni's integral, 121
function, 193
Fundamental Theorem of Calculus, 106–108, 158

G

Gamma function, 121
Gauss's Test, 189
Generalized Rolle's Theorem, 85, 174
geometric mean, 46, 117
geometric series, 125, 135, 136, 163, 168, 169
greatest lower bound, 7

H

Hadamard's formula, 165, 169
harmonic series, 128
Heine-Borel Theorem, 208

I

iff, 191
image, 193
improper integral, 112
    absolutely convergent, 113
    conditionally convergent, 114
induction, 3
    complete, 18
inductive definition, 22
infimum, 11
infinite limit
    of function, 57–58
    of sequence, 29
infinite series.
    absolutely convergent, 42, 137
    conditionally convergent, 42, 139
    convergent, 41, 123
    divergent, 41, 124
    of positive terms, 41–42, 129
    positive and negative parts of, 138
    rearrangement of, 142–144
infinity, 13, 29–30
integrability, 94–98, 102
    of absolute value, 95
    of continuous functions, 102
    of monotone functions, 96
    of products, 98
    of sum, 91
    Riemann's condition for, 94
integral
    alternate definition, 106
    improper, 112
    Riemann, 90
integral test, 129–131
integration by parts, 109
integration by substitution, 108

interior of a set, 202
interior point, 200
intermediate value property, 63–64
Intermediate Value Theorem, 62, 68
    for derivatives, 76
intersection, 192
interval of convergence, 164
inverse function, 193
inverse image, 193
isolated point, 206
iterated limit, 158

K

Kummer's Test, 186

L

Lagrange remainder, 173, 186
least upper bound, 9
Least Upper Bound Theorem, 10, 39
Leibnitz's Test, 140, 142
L'Hospital's Rule, 81–84
limit
    double, 159
    iterated, 158
    left hand, 60
    of function, 49
    of sequence, 23–25
    one sided, 59
    right hand, 59
    sequential criterion for, 53, 55, 58, 60
    uniqueness of, 26, 50
Limit Comparison Test, 132, 133
limit inferior, 42–46
limit superior, 42–46
local maximum, 66
local minimum, 66

M

Mascheroni's constant, 120
maximum, 65
Mean Value Theorem, 77
    Cauchy's, 79–81, 83, 108

    for integrals, 108, 186
minimum, 66
monotone function, 60–61
    limit of, 61
monotone sequence, 31
Monotone Sequence Theorem, 32–34, 39–42, 47
M-test, Weierstrass, 153, 164, 166

N

natural numbers, 2
negative part of a series, 138
neighborhood of a point, 51, 199
Newton's Binomial Theorem, 186
nonintegrable function, 91

O

O, Hardy's big, 188
one-to-one, 193
onto, 193
open cover, 207
open set, 200
    and continuity, 201
oscillation of a function, 68

P

partial sums of a series, 123
partition, 87
    norm, 87
    refinement of, 88
parts,
    integration by, 109
    summation by, 179
pointwise convergence, 149
positive part of a series, 138
power series, 163
    differentiation of, 168
    integration of, 168
product rule, 72

Q

quotient rule, 73

R

Raabe's Test, 187

radius of convergence, 164
range of a function, 193
Ratio Test, 134, 165, 167, 171
rearrangement, 142
Riemann integral, 90
Riemann's Integrability Condition, 94, 96
Riemann sum, 103–106, 107
Rolle's Theorem, 77, 85
    Generalized, 85, 174
Root Test, 136, 165, 167

S
Sandwich Theorem, 28, 44, 54, 105
Second Derivative Test, 79
sequence, 21
    convergent, 24
    decreasing, 31
    divergent, 24
    increasing, 31, 34
    monotone, 31
    strictly decreasing, 31
    strictly increasing, 31
sequential criterion, 53, 55, 58, 60
series, *see* infinite series
set, 191
subcover, 207
subsequence, 35
subset, 192
substitution, integration by, 108
summation by parts, 179
supremum, 9

T
Tauberian Theorem, 183–184
Tauber's Theorem, 182
Taylor polynomial, 173
Taylor series
    convergent to a different function, 172
    divergent, 177
Taylor's Formula, 171
Taylor's Theorem
    integral remainder, 184–185
    Lagrange remainder, 173
telescoping sum, 126
triangle inequality, 7

U
uncountable set, 195, 197
uniform continuity, 99, 102, 209
uniform convergence, 149, 153, 163
    and continuity, 151
    and differentiation, 157
    and integration, 155
union, 192

W
Wallis's Product, 119
Weierstrass M-test, 153, 164, 166
Well-ordering Principle, 3